Graphene and Other 2D Layered Nanomaterial-Based Films

Graphene and Other 2D Layered Nanomaterial-Based Films: Synthesis, Properties and Applications

Special Issue Editors

Federico Cesano
Domenica Scarano

MDPI • Basel • Beijing • Wuhan • Barcelona • Belgrade

MDPI

Special Issue Editors

Federico Cesano
University of Torino
Italy

Domenica Scarano
University of Torino
Italy

Editorial Office
MDPI
St. Alban-Anlage 66
4052 Basel, Switzerland

This is a reprint of articles from the Special Issue published online in the open access journal *Coatings* (ISSN 2079-6412) from 2017 to 2018 (available at: https://www.mdpi.com/journal/coatings/special_issues/Graphene_Film)

For citation purposes, cite each article independently as indicated on the article page online and as indicated below:

LastName, A.A.; LastName, B.B.; LastName, C.C. Article Title. *Journal Name* **Year**, *Article Number*, Page Range.

ISBN 978-3-03921-902-5 (Pbk)
ISBN 978-3-03921-903-2 (PDF)

Cover image courtesy of Federico Cesano.

Contents

About the Special Issue Editors

Federico Cesano, Dr., is a chemist/materials scientist specialized in the nanoscience field with a special focus on surface chemistry and structural properties. In his investigations, he primarily makes use of spectroscopy and electron and scanning probe microscopies, but he also uses other techniques to investigate and characterize structures and materials at the molecular level. He was educated in Italy, obtaining his M.Sc. in Chemistry and receiving his Ph.D. in Materials Science from the University of Torino. He has over fifteen years of experience in the fields of carbon-based materials, oxides, composite/hybrid structures, and their applications. His research interests focus on morphology-structure-property relationships in materials containing 1D, 2D, and 3D nanostructures, including CNTs/graphene and graphene analogues, with particular attention to nanostructured systems and the assembly of nanostructures into functional materials with electrical, magnetic, optical, and/or (photo)catalytic properties. Federico is currently involved in international (EU projects) and national projects with both academic and industrial partners, of which he has an extensive history.

Domenica Scarano, Prof. (born in 1956, Associate Professor since 1999) was Visiting Researcher at the Dept. of Chemical Physics, Fritz Haber Institut in Berlin in 1999. She taught Electrochemistry (1st level in Chemistry Degree), Spectroscopic Methods and Microscopy (1st level Material Science Degree), Interaction and Molecular Recognition (2nd level Chemistry Degree). She currently teaches Physical Chemistry (1st level Chemistry Degree), Spectroscopic Methods and Microscopy, Materials Physical Chemistry (1st level Material Science and Technology Degree) and has supervised numerous theses. Scarano's research has focused on oxide-based materials and, more recently, on hybrid carbon oxide composites as documented by more than 140 ISI publications. She has served as coordinator of numerous research projects. Since 2005, she has been responsible for the PLS in Materials Science of the Torino Unit.

Preface to "Graphene and Other 2D Layered Nanomaterial-Based Films: Synthesis, Properties and Applications"

Graphene, one of the most interesting and versatile materials of the last years, is recognized for its unique properties strongly different from the bulk counterpart. This discovery has stimulated rapid research activity and other two-dimensional (2D) systems, consisting of a single layer of atoms. All of the 2D materials have also emerged among the main candidate materials for many next-generation applications as a result of the considerable and rapid reviews of their properties. In this issue, we have tried to collect a group of papers which examine some of these new areas of work in the field of 2D materials.

Federico Cesano, Domenica Scarano
Special Issue Editors

coatings

MDPI

Editorial

Graphene and Other 2D Layered Hybrid Nanomaterial-Based Films: Synthesis, Properties, and Applications

Federico Cesano * and Domenica Scarano *

Department of Chemistry, NIS (Nanostructured Interfaces and Surfaces) Interdepartmental Centre and INSTM Centro di Riferimento, University of Torino, Via P. Giuria, 7, 10125 Torino, Italy
* Correspondence: federico.cesano@unito.it (F.C.); domenica.scarano@unito.it (D.S.);
  Tel.: +39-11-670-7834 (F.C. & D.S.)

Received: 29 October 2018; Accepted: 21 November 2018; Published: 23 November 2018

Abstract: This Special Issue contains a series of reviews and research articles demonstrating actual perspectives and future trends of 2D-based materials for the generation of functional films, coatings, and hybrid interfaces with controlled morphology and structure.

Keywords: coatings; 2D materials; layered materials; graphene; reduced graphene oxide; transition metal dichalcogenides; WS_2; MoS_2; transition metal carbides; transition metal nitrides; transition metal carbonitrides; silicene; germanene; stanene; van der Waals heterostructures; interfaces

1. Introduction

Graphene is one of the most interesting and versatile materials of the last several years, especially since the Nobel prize in physics was awarded in 2010 to Geim and Novoselov for "groundbreaking experiments regarding the two-dimensional material graphene" [1]. The new material, being "isolated" in a controlled manner and recognised for its unique properties strongly different from those of the bulk counterpart, is a matter of interest for both fundamental studies and practical applications. Whilst the research on graphene has been extremely active since its discovery, a plethora of opportunities has appeared more recently, when other 2D layered systems and their combinations (i.e., van der Waals heterostructures) have been taken into consideration [2]. Moreover, two-dimensional (2D) systems, consisting of a single layer of atoms, have emerged among the main candidate materials for next-generation applications [3–5]. In general terms, the strict limit of the one atomic layer in thickness of these 2D crystals does not matter when new properties and applications with respect to 3D counterparts are taken into account. Accordingly, a material exhibiting some unique properties is, in fact, still considered a 2D material even if it is made of one/two/three or more layers. In such cases, they are described as being of monolayer, bilayer, three-layer, or few-layer thickness, but these materials have the potential to revolutionise electronics concepts and make new technologies feasible.

At the time of writing this Special Issue, a few dozen materials made of crystalline and one-atom-thick systems have been successfully obtained by exfoliation of 3D compounds (top-down approach) or by synthetic procedures (bottom-up approach) [6], but it is hard to give a more precise number of the 2D crystals due to the fast advancement in the field. Further, due to the discovery in 2017 of magnetic 2D materials, rapid progress in this field can also be mentioned. On this matter, significant examples can be highlighted, including magnetic single-layer CrI_3 (i.e., odd layer numbers, the magnetisation being absent for an even number of layers due to the antiferromagnetic coupling between the layers) [7] and ferromagnetic two-layered $Cr_2Ge_2Te_6$ [8]. Notice that, although all 2D materials are expected to be inorganics, chromium–chloride–pyrazine ($CrCl_2(pyrazine)_2$) is the first discovered organic/inorganic hybrid 2D material [9]. Together with its other prominent properties, 2D $CrCl_2(pyrazine)_2$ exhibits magnetic properties.

2. 2D Materials: Quō Vādis?

Recently, Mounet et al. [10] showed that only a very small fraction of possible 2D crystals—belonging to transition metal carbides, nitrides, or carbonitrides (MXenes) [11]; silicene, germanene, or stanene (Xenes) [12]; transition metal dichalcogenides (MX_2) [6]; and graphene and graphene derivatives [13]—have been considered so far. Therefore, most 2D materials have not yet been discovered. In this regard, nothing can be said to be certain about the next one-atom-thick material. However, some possible highlights can be envisaged, including more simple fabrication techniques [14]; precise control of size and shape; greener production methods; 2D crystal doping [15]; superconducting properties of 2D crystals [16]; atom-by-atom assembling of 2D materials directly onto the surface of solids, such as photoactive TiO_2 polytypes [17]; or an energy breakthrough of 2D crystals (i.e., graphene) as a source of clean, limitless energy due to the layer motion (e.g., rippled morphology and temperature-induced curvature inversion) [18].

3. This Special Issue

This Special Issue, entitled "Graphene and Other 2D Layered Nanomaterial-Based Films: Synthesis, Properties, and Applications", contains a collection of three reviews and eight research articles covering fundamental studies and applications of films and coatings based on 2D materials. Going into detail, the thermal growth of graphene and the advances in the field of free-standing graphene films for thermal applications are comprehensively reviewed by Tan et al. [19] and Gong et al. [20], respectively. The first review focuses on the mechanisms and main fabrication methods (epitaxial growth, chemical vapour deposition, plasma-enhanced chemical vapour deposition, and combustion), summarising the latest research progress in optimising growth parameters. Besides synthesis methods, the second review is dedicated to interface properties and the thermal conductivity of materials based on free-standing graphene nanosheets, as well as their thermal applications (e.g., heat dissipation materials, wearable flexible materials for thermal control). Along with surface-enhanced Raman spectroscopy (SERS), 2D-material-coated plasmonic structures are described in the review article by Xia [21]. In this review, the effects and advantages of combining 2D materials with traditional metallic plasmonic structures (i.e., higher SERS enhancement factors, oxidation protection of the metal surface, and protection of molecules from photo-induced damage) have been highlighted.

The preparation, properties, and applications of some 2D materials (i.e., graphene, graphene oxide, WS_2, MoS_2, and 2D carbon nitride nanosheets in a nickel–phosphorus alloy) are discussed in the eight research papers. Briefly, the fundamental work conducted by Lee et al. [22] provides a valuable insight into the nondestructive transfer of graphene from the surface of a metal catalyst to target substrates, without dissolving the metallic catalyst by chemical etching. Tsai et al. [23] report the preparation of a graphene-coated electrode by a spin-coating technique and the consequent effect on enhancing bacterial adhesion and increasing the power generation of the deposited film in microbial fuel cells (MFCs). Lv and Zhao et al. have investigated the preparation by a chemical vapour deposition (CVD) technique and photoluminescence properties of a WS_2 monolayer (which is a direct bandgap semiconductor) [24] in a first article and the preparation of mono- and few-layered MoS_2 by a CVD technique using water as a transport agent and growth promoter of the MoS_2 sheets [25] in a second paper. Mardle et al. [26] have evaluated the catalytic power performance of aligned Pt nanowires grown on reduced graphene oxide in proton-exchange membrane fuel cell (PEMFC) electrodes, while MoS_2 nanosheets supported on Pt nanoparticles have been obtained by Cheng et al. [27] to enhance the power conversion efficiency (PCE) of dye-sensitised solar cells (DSSCs) up to 7.52%. Alternatively, Shi et al. [28] have grown graphene/few-layered MoS_2/Si heterostructures by a CVD technique, and they investigated the double-junction properties in terms of enhancing the photovoltaic performance of van der Waals heterostructures. Finally, Fayyad et al. [29] have obtained 2D carbon nitride (C_3N_4) nanosheets in a nickel–phosphorus (NiP) matrix by ultrasonication during electroless plating of NiP. The microhardness and corrosion resistance of the as-modified coatings have been evaluated and compared with those of the native NiP alloy.

In summary, this Special Issue of *Coatings* compiles a series of reviews and research articles demonstrating the potential of 2D-based materials for the generation of functional films, coatings, and hybrid interfaces with controlled morphology and structure.

Conflicts of Interest: The authors declare no conflict of interest.

References

1. The Nobel Prize in Physics 2010. Available online: http://www.nobelprize.org/nobel_prizes/physics/laureates/2010/ (accessed on 15 November 2018).
2. Novoselov, K.S.; Mishchenko, A.; Carvalho, A.; Castro Neto, A.H. 2D materials and van der Waals heterostructures. *Science* **2016**, *353*, aac9439. [CrossRef] [PubMed]
3. Zeng, M.; Xiao, Y.; Liu, J.; Yang, K.; Fu, L. Exploring two-dimensional materials toward the next-generation circuits: From monomer design to assembly control. *Chem. Rev.* **2018**, *118*, 6236–6296. [CrossRef] [PubMed]
4. Sun, Y.; Chen, D.; Liang, Z. Two-dimensional MXenes for energy storage and conversion applications. *Mater. Today Energy* **2017**, *5*, 22–36. [CrossRef]
5. Roldan, R.; Chirolli, L.; Prada, E.; Silva-Guillen, J.A.; San-Jose, P.; Guinea, F. Theory of 2D crystals: Graphene and beyond. *Chem. Soc. Rev.* **2017**, *46*, 4387–4399. [CrossRef] [PubMed]
6. Manzeli, S.; Ovchinnikov, D.; Pasquier, D.; Yazyev, O.V.; Kis, A. 2D transition metal dichalcogenides. *Nat. Rev. Mater.* **2017**, *2*, 17033. [CrossRef]
7. Huang, B.; Clark, G.; Navarro-Moratalla, E.; Klein, D.R.; Cheng, R.; Seyler, K.L.; Zhong, D.; Schmidgall, E.; McGuire, M.A.; Cobden, D.H.; et al. Layer-dependent ferromagnetism in a van der Waals crystal down to the monolayer limit. *Nature* **2017**, *546*, 270–273. [CrossRef] [PubMed]
8. Gong, C.; Li, L.; Li, Z.; Ji, H.; Stern, A.; Xia, Y.; Cao, T.; Bao, W.; Wang, C.; Wang, Y.; et al. Discovery of intrinsic ferromagnetism in two-dimensional van der Waals crystals. *Nature* **2017**, *546*, 265–269. [CrossRef] [PubMed]
9. Pedersen, K.S.; Perlepe, P.; Aubrey, M.L.; Woodruff, D.N.; Reyes-Lillo, S.E.; Reinholdt, A.; Voigt, L.; Li, Z.; Borup, K.; Rouzières, M.; et al. Formation of the layered conductive magnet $CrCl_2$(pyrazine)$_2$ through redox-active coordination chemistry. *Nat. Chem.* **2018**, *10*, 1056–1061. [CrossRef] [PubMed]
10. Mounet, N.; Gibertini, M.; Schwaller, P.; Campi, D.; Merkys, A.; Marrazzo, A.; Sohier, T.; Castelli, I.E.; Cepellotti, A.; Pizzi, G.; et al. Two-dimensional materials from high-throughput computational exfoliation of experimentally known compounds. *Nat. Nanotechnol.* **2018**, *13*, 246–252. [CrossRef] [PubMed]
11. Hong Ng, V.M.; Huang, H.; Zhou, K.; Lee, P.S.; Que, W.; Xu, J.Z.; Kong, L.B. Recent progress in layered transition metal carbides and/or nitrides (MXenes) and their composites: Synthesis and applications. *J. Mater. Chem. A* **2017**, *5*, 3039–3068. [CrossRef]
12. Molle, A.; Goldberger, J.; Houssa, M.; Xu, Y.; Zhang, S.C.; Akinwande, D. Buckled two-dimensional Xene sheets. *Nat. Mater.* **2017**, *16*, 163–169. [CrossRef] [PubMed]
13. Inagaki, M.; Kang, F. Graphene derivatives: Graphane, fluorographene, graphene oxide, graphyne and graphdiyne. *J. Mater. Chem. A* **2014**, *2*, 13193–13206. [CrossRef]
14. Shim, J.; Bae, S.H.; Kong, W.; Lee, D.; Qiao, K.; Nezich, D.; Park, Y.J.; Zhao, R.; Sundaram, S.; Li, X.; et al. Controlled crack propagation for atomic precision handling of wafer-scale two-dimensional materials. *Science* **2018**, *342*, 833–836. [CrossRef] [PubMed]
15. Feng, S.; Lin, Z.; Gan, X.; Lv, R.; Terrones, M. Doping two-dimensional materials: Ultra-sensitive sensors, band gap tuning and ferromagnetic monolayers. *Nanoscale Horiz.* **2017**, *2*, 72–80. [CrossRef]
16. Saito, Y.; Nojima, T.; Iwasa, Y. Highly crystalline 2D superconductors. *Nat. Rev. Mater.* **2016**, *2*, 16094. [CrossRef]
17. Cravanzola, S.; Cesano, F.; Gaziano, F.; Scarano, D. Carbon domains on MoS_2/TiO_2 system via acetylene oligomerization: Synthesis, structure and surface properties. *Front. Chem.* **2017**, *5*, 91. [CrossRef] [PubMed]

18. Ackerman, M.L.; Kumar, P.; Neek-Amal, M.; Thibado, P.M.; Peeters, F.M.; Singh, S. Anomalous dynamical behavior of freestanding graphene membranes. *Phys. Rev. Lett.* **2016**, *117*, 126801. [CrossRef] [PubMed]

19. Tan, H.; Wang, D.; Guo, Y. Thermal growth of graphene: A review. *Coatings* **2018**, *8*, 40. [CrossRef]

20. Gong, F.; Li, H.; Wang, W.; Xia, D.; Liu, Q.; Papavassiliou, D.V.; Xu, Z. Recent advances in graphene-based free-standing films for thermal management: Synthesis, properties, and applications. *Coatings* **2018**, *8*, 63. [CrossRef]

21. Xia, M. 2D materials-coated plasmonic structures for SERS applications. *Coatings* **2018**, *8*, 137. [CrossRef]

22. Lee, J.; Lee, S.; Yu, H.K. Contamination-free graphene transfer from Cu-foil and Cu-thin-film/sapphire. *Coatings* **2017**, *7*, 218. [CrossRef]

23. Tsai, H.-Y.; Hsu, W.-H.; Liao, Y.-J. Effect of electrode coating with graphene suspension on power generation of microbial fuel cells. *Coatings* **2018**, *8*, 243. [CrossRef]

24. Lv, Y.; Huang, F.; Zhang, L.; Weng, J.; Zhao, S.; Ji, Z. Preparation and photoluminescence of tungsten disulfide monolayer. *Coatings* **2018**, *8*, 205. [CrossRef]

25. Zhao, S.; Weng, J.; Jin, S.; Lv, Y.; Ji, Z. Chemical vapor transport deposition of molybdenum disulfide layers using H_2O vapor as the transport agent. *Coatings* **2018**, *8*, 78. [CrossRef]

26. Mardle, P.; Fernihough, O.; Du, S. Evaluation of the scaffolding effect of pt nanowires supported on reduced graphene oxide in PEMFC electrodes. *Coatings* **2018**, *8*, 48. [CrossRef]

27. Cheng, C.-K.; Lin, J.-Y.; Huang, K.-C.; Yeh, T.-K.; Hsieh, C.-K. Enhanced efficiency of dye-sensitized solar counter electrodes consisting of two-dimensional nanostructural molybdenum disulfide nanosheets supported Pt nanoparticles. *Coatings* **2017**, *7*, 167. [CrossRef]

28. Shi, W.; Ma, X. Photovoltaic effect in graphene/MoS_2/Si van der Waals heterostructures. *Coatings* **2017**, *8*, 2. [CrossRef]

29. Fayyad, E.M.; Abdullah, A.M.; Hassan, M.K.; Mohamed, A.M.; Wang, C.; Jarjoura, G.; Farhat, Z. Synthesis, characterization, and application of novel Ni-p-carbon nitride nanocomposites. *Coatings* **2018**, *8*, 37. [CrossRef]

coatings

Review

Thermal Growth of Graphene: A Review

Hai Tan [1], Deguo Wang [1,2] and Yanbao Guo [1,2,*]

[1] College of Mechanical and Transportation Engineering, China University of Petroleum,
    Beijing 102249, China; doc.tan@outlook.com (H.T.); wdg@cup.edu.cn (D.W.)
[2] Beijing Key Laboratory of Fluid Filtration and Separation, China University of Petroleum,
    Beijing 102249, China
*   Correspondence: gyb@cup.edu.cn; Tel.: +86-10-8973-3727

Received: 29 November 2017; Accepted: 30 December 2017; Published: 19 January 2018

Abstract: A common belief proposed by Peierls and Landau that two-dimensional material cannot exist freely in a three-dimensional world has been proved false when graphene was first synthesized in 2004. Graphene, which is the base structure of other carbon materials, has drawn much attention of scholars and researchers due to its extraordinary electrical, mechanical and thermal properties. Moreover, methods for its synthesis have developed greatly in recent years. This review focuses on the mechanism of the thermal growth method and the different synthesis methods, where epitaxial growth, chemical vapor deposition, plasma-enhanced chemical vapor deposition and combustion are discussed in detail based on this mechanism. Meanwhile, to improve the quality and control the number of graphene layers, the latest research progress in optimizing growth parameters and developmental technologies has been summarized. The strategies for synthesizing high-quality and large-scale graphene are proposed and an outlook on the future synthesis direction is also provided.

Keywords: graphene; epitaxial growth; chemical vapor deposition; plasma; combustion

---

1. Introduction

The wide knowledge that a strictly two-dimensional crystal cannot exist was disproved when graphene was first isolated by Geim and Novoselov at the University of Manchester in 2004 [1–4]. Thus, the carbon family consists of each dimensional material: fullerene in zero dimensions [5]; carbon nanotube in one dimension [6]; graphene in two dimensions (2D); and graphite in three dimensions (3D). Graphene, a one-atom thick layer of sp^2 hybridized carbon atoms arranged into hexagonal crystal, has been a topic of interest in nano-science due to its excellent properties and the prospect of industrial applications [7–10]. Owning to its unique structure, the charge carrier mobility of graphene exceeds 2.0×10^5 cm$^2 \cdot$V$^{-1} \cdot$s^{-1} at room temperature which is 100 times higher than that of silicon [11]. Moreover, graphene is one of the strongest materials in the world and its Young's modulus is more than 1 TPa [12]. Graphene also shows a good thermal conductivity of 5000 W$\cdot$mK^{-1} and optical performance with an opacity of 2.3% per layer [13,14]. However, obtaining graphene with high quality and large scale is still a difficult problem to solve.

Since the "scotch tape method" [4] which helps to study the properties of graphene, various kinds of strategies have been developed to synthesize this 2D carbon material. These methods could be divided into "top-down" stripping methods and "bottom-up" synthesis methods. The stripping method consists of peeling the stacked graphene sheet from graphite through external force, such as normal stress and sheer stress. When the external force is bigger than the Van der Waals' force between the molecular layers, graphene can be peeled (see Figure 1) [15]. Conversely, the synthesis method relies on the recombination of carbon atoms. The stripping method mainly comprises of mechanical cleavage and the oxidation-reduction method. Although graphene achieved by mechanical cleavage method has better quality and is an easier manufacturing technique, the product only just meets the experimental requirement. The oxidation-reduction method can produce graphene with high yield, however the graphene always has many structure defects. The synthesis method, such as chemical vapor deposition (CVD) and epitaxial growth, can output high-quality and large-scale graphene. Moreover, graphene achieved in this way meets the needs of the electronic and optoelectronic industries [16,17].

Figure 1. Mechanism of stripping method.

Chemical vapor deposition and epitaxial growth are not economic. However, with the improvement of production process, synthesizing high-quality and large-scale graphene at low cost is possible. The thermal growth method, as one of the synthesis approaches, has been widely discussed before. This review provides the research progress of graphene production, studying not only the thermal growth technology itself, but also the thermal growth mechanism in detail. Furthermore, the conclusion of the thermal growth method and the development prospects for producing high-quality and large-scale graphene at low cost are introduced.

2. Thermal Method for Growing of Graphene

The thermal method for growing graphene has the potential to produce high-quality and large-scale graphene compared to the stripping method. The thermal method is always high yield and meets requirements of various industries. However, it is expensive, and more complicated equipment is often needed. The difficult transfer process and high temperature also constrain the development of the thermal method. Hence, if we want to get high-quality and large-scale graphene with high benefits, this growth process should be well understood. The mechanism of the thermal method is shown in Figure 2. Carbon atoms always link with other atoms in different chemical bonds, such as sp^3 bonds. In order to achieve graphene, individual carbon atoms should be released initially through exerting extra energy, and then they nucleate with others in the structure of benzene ring through sp^2 bonds. After that, the nucleation grows into graphene. In brief, the mechanism of the thermal growth method is the split of molecules and recombination of atoms. It should be stated that this mechanism is adapted to the thermal method for growing of graphene illustrated later, and the only difference is the way of destroying the molecular bond. based on this growth process, three main methods to synthesize graphene are proposed and summarized in Table 1.

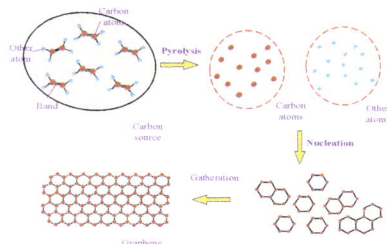

Figure 2. Mechanism of thermal method for growing of graphene.

Table 1. A summary of three different methods to synthesize graphene.

Method		Advantage	Disadvantage
Epitaxial growth		High quality; highly compatible with electronics	High costs of SiC wafers; Low yield; Hard to transfer
Chemical vapor deposition	Conventional chemical vapor deposition	Large graphene films; Possible to transfer onto multitudes of materials; High quality and large-scale production	Required substrates are often expensive; Complicated synthetic and transfer process; Introducing new defects in the transfer process
	Plasma-enhanced chemical vapor deposition	Relative low temperature; Short reaction time	
Combustion method		Simple facility; Quick synthetic process;	Hard to control the combustible process; Non-uniform distribution; Low quality

2.1. Epitaxial Growth of Graphene

It was reported in 1962 [18] that when silicon carbide (SiC) is heated to a certain temperature, the silicon carbide shows graphitization and the product always contains amorphous carbon and multilayer graphite. With the development of the epitaxial technique, graphene can be achieved while putting the etching SiC substrate into a high temperature and ultra-high vacuum vessel for a relatively long time. Figure 3 shows the theory of epitaxial growth of graphene. It can be observed that carbide decomposes in the experimental process, and then the carbon atom recombines while non-carbon atoms evaporate. This method is almost based on the SiC substrate, thus the products have a good compatibility with integrated circuits. 6H–SiC and 4H–SiC are often selected to act as the original carbon sources, because both of them have the same Si-C bilayer structure [19,20]. Table 2 is a summary of the epitaxial growth method and the main properties of the synthesized graphene.

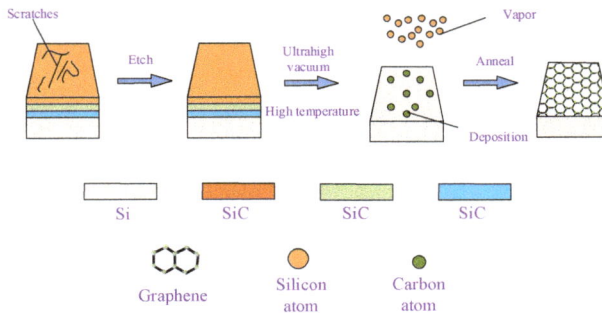

Figure 3. Mechanism of epitaxial growth of graphene.

Table 2. A summary of epitaxial growth method and the properties of the synthesized graphene.

Substrate	Precursor Gas	Pressure (Torr)	Temperature (°C)	Charge Carrier Mobility (cm$^2 \cdot$V$^{-1} \cdot$s^{-1})	Square Resistance (kΩ/sq)	Ref.
6H–SiC	–	1×10^{-10}	1450	1100 (4 K)	1.5 (4 K)	[21]
4H–SiC	–	ultra-high vacuum	–	2.5×10^4 (2490 K)	1.41 (30 K)	[22]
Ni/6H–SiC	–	4.5×10^{-10}	950	–	–	[23]
6H–SiC	–	4.5×10^{-10}	1300	–	–	[24]
6H–SiC	Argon	750	1550	2000 (27 K)	–	[25]

Beger and his team [21,26] found that the ultrathin graphene can be synthesized on the surface of 6H–SiC in ultra-high vacuum with about 1×10^{-10} Torr and high temperature that changed from 1250 °C to 1450 °C. The low-energy electron diffraction (LEED) pattern was used to characterize the different growth states of graphene in situ, as shown in Figure 4. It can be seen that with the increase in temperature, the SiC first changes from 1×1 pattern to $\sqrt{3} \times \sqrt{3}$ transition structure, and then a $6\sqrt{3}$

$\times\ 6\sqrt{3}$ unit cell is achieved. Finally, the graphene with charge carrier mobility 1100 cm^2·V^{-1}·s^{-1} at 4 K is achieved.

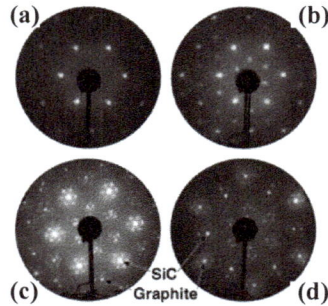

Figure 4. LEED patterns in different temperatures and times (Reproduced from [21] with permission; Copyright 2004 American Chemical Society). (**a**) 1050 °C, 10 min; (**b**) 1100 °C, 3 min; (**c**) 1250 °C, 20 min; (**d**) 1400 °C, 8 min.

In order to control the quality of produced graphene, plenty of researchers are dedicated to various kinds of studies on epitaxial growth, such as the investigations of experimental parameters and detection means [23,27,28]. The quality of produced graphene in the ultra-high vacuum is hard to master and has more defects. Meanwhile, excessively high or low temperature also leads to the reduction of graphene quality. When the temperature is excessively high, the number of graphene layers increases. The reflective high energy electron diffraction (RHEED) and the atomic force microscope (AFM) were chosen to study the influence of annealing time, and the results showed that the number of graphene layers is related to the annealing time [24]. The growth pressure was well-controlled by introducing argon (Ar) as a buffer gas into the experimental environment, and the growth mechanism was also discussed by Seyller et al. [25]. They found the charge carrier mobility of the obtained product can reach 2000 cm^2·V^{-1}·s^{-1} at 27 K and explained that the Ar could not only decrease the growth rate and guarantee the growth temperature, but also decrease the vapor rate of silicon atoms.

The graphene could be also synthesized by another carbide, such as titanium carbide (TiC) [29] and tantalum carbide (TaC) [30]. However, these carbides are little studied due to the needs of particular crystal structures and far higher experimental temperature. With the development of other 2D material, hexagonal boron nitride (h-BN) [31] is also regarded as a substrate to epitaxial growth, which is a new idea for further research. The expensive materials and complicated transfer process limit the mass production of graphene. Thus, more attention should be paid to the new carbide than the existing materials, or the manufacturing technique should be changed to maximize profits.

2.2. Chemical Vapor Deposition Mechanism

Chemical vapor deposition (CVD) has the potential to synthesize high-quality graphene that can satisfy the needs of industry. Table 3 shows the typical graphene properties for various kinds of chemical vapor deposition.

Table 3. Typical graphene properties for various kinds of chemical vapor deposition.

Method		Substrate	Precursor Gas	Temperature (°C)	Number of Layer	Size (cm^2)	Ref.
Conventional chemical vapor deposition		Ni	CH_4; H_2	900; 1000	1–12	2	[32]
			CH_4; H_2; Ar	1000	1–10	4	[33]
			Soybean	800	–	4	[34]
		Cu	CH_4; H_2	1000	1–3	1	[35]
			CH_4; H_2	1000	1	30 (inch)	[36]
			Polystyrene; H_2; Ar	1000	1	1	[37]
Plasma-enhanced chemical vapor deposition	Micro-wave-assisted	Various	CH_4; H_2	700	4–6	–	[38]
		Cu	CH_4; H_2	<420	1	1.04	[39]
		Non	C_2H_5OH; Ar	–	–	–	[40]
	Arc-discharge	–	H_2; graphite; Ar	–	2–4	–	[41]

Various kinds of materials can be used as substrate to synthesize graphene. Traditional materials, such as copper (Cu) foils and nickel (Ni) are widely employed. Most recently, researchers have paid much attention to other 2D materials, including h-BN [42–44] and molybdenum disulfide (MoS$_2$) [45,46]. Cu and Ni can be seen as the representative of low carbon soluble and high carbon soluble materials, respectively. The mechanism to grow graphene on Cu is illustrated in Figure 5. Figure 5a,b show that the surface is etched by hydrogen (H$_2$) at high temperature until there are no obvious scratches. After that, the carbon source and buffer gas are introduced into the reactive system (Figure 5c,d). When the carbon source contacts the Cu at the high temperature, it dissociates into atoms and carbon atoms deposit on the Cu surface. However, because of the low carbon solubility in Cu with 0.008 wt % in 1084 °C [47], the carbon atoms will not further permeate into the Cu. These deposited carbon atoms combine with others to form "graphene islands"; the islands enlarge and further unite to graphene, shown in Figure 5d1,d2.

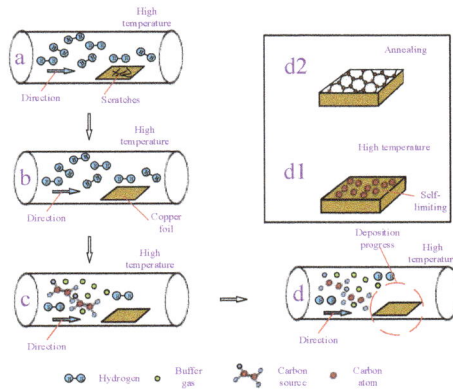

Figure 5. Mechanism of CVD growth graphene on Cu: (**a**) before etching; (**b**) after etching by hydrogen; (**c**) introducing buffer gas and carbon source; (**d**) synthesis process; (**d1**) before annealing; (**d2**) after annealing.

The high-quality carbon nanotube was grown on the Ni surface using CVD [48]. It was not until 2009 that graphene synthesized on Ni by CVD through improving the experimental parameters and conditions was reported [32,33]. The mechanism is similar to that of growth on Cu, and the difference is shown in Figure 6. Graphene growth on Cu is mainly dependent on self-limiting; however, for Ni it is mainly caused by separating out of the carbon atoms due to the relatively high carbon solubility in Ni with 0.6 wt % at 1326 °C [47]. It is clear that a carbon source decomposes at high temperature when it contacts the Ni surface. After that, the splitting carbon atoms permeate into Ni to form a solid solution with Ni. Finally, the supersaturated carbon atoms separate out and synthesize graphene on the Ni surface after annealing.

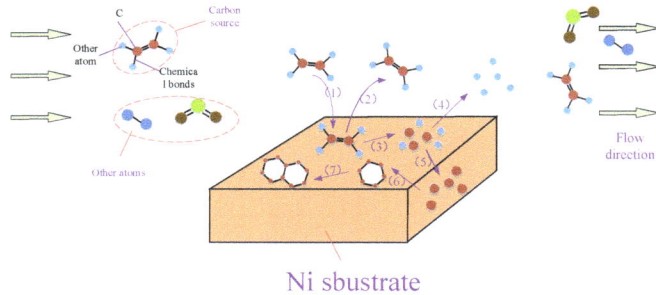

Figure 6. Mechanism of CVD growth graphene on Ni.

2.2.1. Conventional Chemical Vapor Deposition Growth on Ni

Researchers at the Massachusetts Institute of Technology have systematically studied the growth of graphene on the polycrystalline Ni substrate. Two temperatures, 900 °C and 1000 °C, were studied in different concentrations of methane (CH_4) and H_2. In addition, the Ni surface was analyzed by transmission electron microscopy (TEM) [32]. As shown in Figure 7, it is clear that the deposited products are graphene and the number of films mainly range from 1 to 8. In subsequent experiments, the coverage of about 87% of monolayer and bilayer graphene were achieved in the ambient pressure by controlling the CH_4 concentration and cooling speed of polycrystalline Ni substrate [49]. Ar was also introduced into the CVD method as a buffer gas and a cooling gas [33]. The 300 nm thick nickel layer on the Si wafer was put in the atmosphere of mixed gases (Ar, H_2 and CH_4) and high temperature of 1000 °C, then the substrate was quickly cooled down to about 25 °C. The number of obtained graphene layers was less than 10. They found that the quality of produced graphene is positively related to the cooling rate. Optical image and Raman spectrum were used to identify the film after being transferred to SiC, as shown in Figure 8. The typical feature for graphene in Raman spectrum is characterized with a G peak (~1580 cm^{-1}), D peak (~1345 cm^{-1}) and 2D peak (~2700 cm^{-1}). In addition, the number of graphene films can be deduced by the ratio of the intensities of 2D peak and G peak [32].

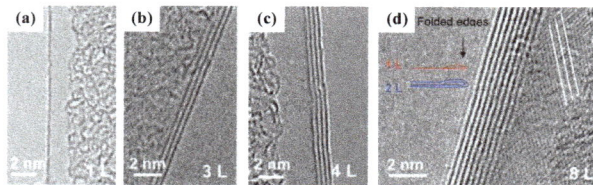

Figure 7. TEM images of graphene films grown on the Ni substrate: (**a**) 1 layer, (**b**) 3 layers, (**c**) 4 layers, (**d**) 8 layers. (Reproduced from [32] with permission; Copyright 2009 American Chemical Society.)

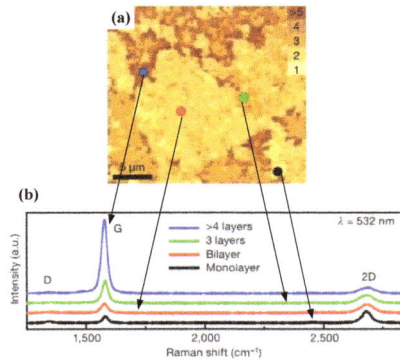

Figure 8. Optical image (**a**) and Raman spectrum (**b**) of graphene films growth on the 300 nm-thick-nickel layer on the Si wafer. (Reproduced from [33] with permission; Copyright 2009 Nature.)

Other parameters have also been discussed in recent years. The difference of grain of Ni was reported by Zhou et al. [50]. They found that because of the atomically smooth surface of single crystalline Ni, the area of monolayer or bilayer graphene deposited on single crystalline Ni (91.4%) is much bigger than that for polycrystalline Ni (72.8%) [50]. Choi et al. [51] systematically studied the mechanism of wrinkle formation and provided optimized parameters to synthesize graphene on single-crystalline Ni through changing the proportion of precursor gases, experimental temperature and deposited time. In order to decrease the cost, various kinds of carbon sources, such as soya-bean oil [34], have been presented.

2.2.2. Conventional Chemical Vapor Deposition Growth on Cu

It is possible to produce high-quality graphene with fewer layers using Cu foil as substrate. Ruoff et al. [35] have synthesized monolayer graphene on the 25 μm thick Cu foil in high temperature conditions (1000 °C) and mixture gases (CH_4 and H_2). The number of graphene layers is lower than 3, and the area of monolayer graphene is more than 95%.

Since then, various research has discussed the effect of experimental parameters to deposit uniform graphene films on the Cu foils, including experimental pressure, time and carbon sources. At a pressure of 340 Pa, graphene is grown on the surface of Cu foil in different times, i.e., 20 min, 80 min, 120 min and 180 min [52]. The optical images in Figure 9 show that with the increase in experimental time, the substrate surface becomes much smoother, thus the growth of bilayer graphene is suppressed and large-scale monolayer graphene is achieved. [46]. Vlassiouk and Smirnov [53] paid much attention on growth temperature and pointed out that the experimental temperature, which is similar to the melting point of Cu foil, contributes to the synthesis of high-quality graphene in the atmospheric pressure. They [53] explained that the sublimation of Cu atoms are restrained and more energy is used for desorption of carbon clusters in that temperature. The influences of growth temperature, CH_4 flow rate and annealing temperature were intensively studied to synthesize high-quality graphene at ambient pressure. The results showed that higher annealing temperature and lower growth temperature contribute to the synthesis of monolayer graphene [54]. In order to obtain graphene at a large scale, roll-to-roll production technology was introduced into the CVD system. In addition, graphene in one 30-inch layer was produced on the ultra-large copper substrate, which can be seen in Figure 10 [36]. Polystyrene was chosen to produce carbon atoms, and monolayer graphene with a coverage of 97.5% on the substrate was achieved in a short time (about 3 min) [37].

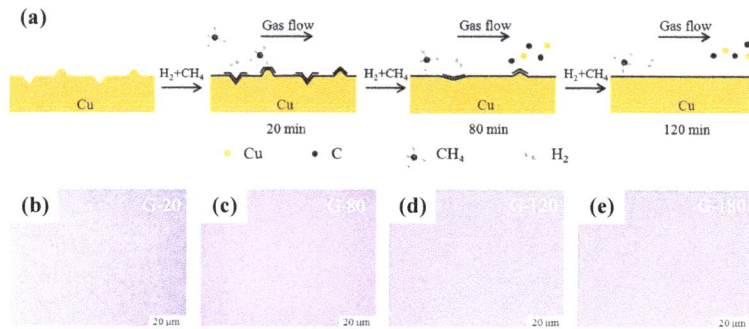

Figure 9. Mechanism of graphene deposition on the Cu in different times and its corresponding optical images. (**a**) Mechanism illustration; (**b**) Optical image after growing 20 min; (**c**) Optical image after growing 80 min; (**d**) Optical image after growing 120 min; (**e**) Optical image after growing 180 min. (Reproduced from [52] with permission; Copyright 2014 Elsevier.)

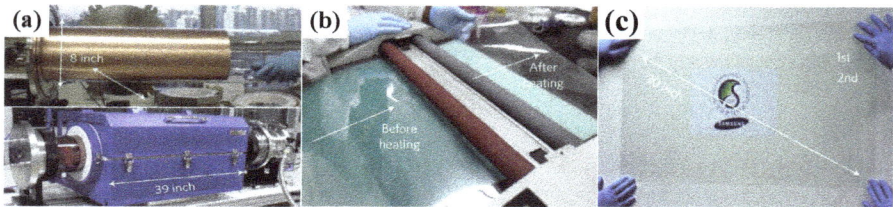

Figure 10. Synthesized progress when introducing roll-to-roll technology into CVD. (**a**) CVD; (**b**) Roll-to-toll technology; (**c**) Graphene with 30-inch. (Reproduced from [36] with permission; Copyright 2010 Nature.)

Meanwhile, researchers focused on the measurement means for exploring the growth progress to produce high-quality graphene [55–58]. For example, the isotope-labelling method and nano angle resolved photoelectron spectroscope (Nano-ARPES) were employed by Ruoff et al. [55] and Asensio et al. [57], respectively. Different substrates, including platinum (Pt) [59], iridium (Ir) [60] and germanium (Ge) [61,62] have also been mentioned. It was first reported that the graphene deposited on the Co/MgO substrate has better application prospects for molecular electronics and polymer composites [63]. To reduce energy consumption and deposit graphene directly on the electronic device, using low temperature to synthesize graphene has become a research hot spot. Various kinds of CVD methods, including hydrogen-free chemical vapor deposition (HFCVD) [64], carbon-enclosed chemical vapor deposition (CECVD) [65], plasma enhanced chemical vapor deposition (PECVD) [66], ultra-high vacuum chemical vapor deposition (UHVCVD) [67] and oxygen-free chemical vapor deposition (OFCVD) [68] method, are presented. Cho and Hong et al. [64] tried to decrease the experimental temperature, however the temperature only reduced from 1000 °C to 970 °C. Jang et al. [68] developed an oxygen-free chemical method which removed the oxygen and successfully used low activation energy benzene as carbon source to synthesize graphene on Cu foils at atmospheric pressure at 300 °C.

2.2.3. Plasma-Enhanced Chemical Vapor Deposition

Graphene can be achieved by conventional chemical vapor deposition, and different CVD methods are explored that have been previously mentioned. Based on the development of plasma technology and requirements of low energy [69], the plasma-enhanced chemical vapor deposition method has been studied [70,71].

In the microwave-assisted deposition progress, high-energy electrons can provide enough activation energy. Once the plasma gases collide with carbon sources, bands of reactive gas are broken and the chemical activity of precursor gases increases. Thus, the experimental temperature decreases. Following this, some atoms recombine with other atoms before coming into contact with the substrate, and some carbon atoms may deposit on the surface of the substrate or permeate into the substrate. After annealing, graphene is synthesized on the substrate surface.

Graphene has been successfully synthesized by microwave-assisted deposition technology at low experimental temperature. For example, a few layers of graphene can be synthesized in the phenomenon of CH_4 and H_2 when the temperature is 700 °C [38]. The synthesized graphene is vertically grown on the substrates, and the number of obtained graphene layers are mainly within the range of 4 to 6 [38]. Moreover, Boyd et al. [39] used Cu foil as a substrate to grow graphene in a relative low temperature (below 420 °C) and the deposition time was just a few minutes. Just as with conventional chemical vapor deposition, different materials that are used as carbon sources are discussed in the studies. Tatarova et al. [72] chose ethyl alcohol as a carbon source to deposit graphene on the surface of a cavity wall and found that the contents of monolayer graphene decrease when the wall's temperature increases from 60 to 100 °C. In addition, when the temperature is 60 °C, the ordered crystal edges are observed. However, with the temperature increasing to 80 °C, the contents of monolayer graphene are clearly reducing. Furthermore, researchers have also paid much attention to various kinds of growth substrates. For example, Song et al. [73] attempted to grow a few graphene layers on metal and nonmetal substrates at low temperature.

The size of graphene limits its further application, and large-scale graphene is desperately needed in many industries. Yamada et al. [74] combined roll-to-roll technology and microwave-assisted chemical vapor deposition technology to build a new graphene preparation system. The mixed gases (CH_4, H_2 and Ar) flowed into this system, and graphene film with dimensions 294 mm × 480 mm was achieved at 400 °C. Dato at the University of California, Berkeley proved [40,75,76] that graphene can be synthesized without substrate through presenting a novel microwave-assisted equipment, as shown in Figure 11. Ethanol droplets flowed through the Ar plasma and graphene was collected in the upper nylon filter. The deposition velocity reached about 2 mg/min. They also found that this method is not fit for CH_4 and isopropanol, and lower velocity contributes to the synthesis graphite instead of graphene.

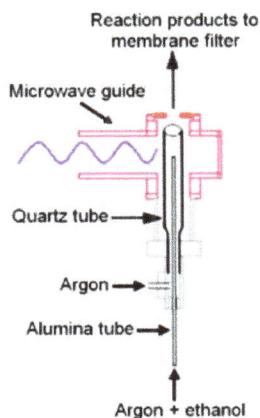

Figure 11. Schematic of a novel microwave-assisted equipment. (Reproduced from [40] with permission; Copyright 2008 American Chemical Society.)

The arc discharge method is another plasma technology that attracts researchers' attentions. Two graphite electrodes are regarded as the carbon source and arc plasma is generated between the two graphite electrodes in the presence of Ar, helium (He) or H_2. Finally, graphene is deposited on the cathodic electrode, reactor wall or special substrates. Nowadays, graphene is successfully synthesized by arc plasma technology, but the control progress should be improved. Rao et al. [41] used direct current arc to vapor anode graphite rod in the phenomenon of H_2, and the number of produced graphene layers was less than 4. They also found that the present of H_2 suppresses the product, changing from graphene to carbon nanotubes. In order to conveniently distinguish the carbon nanotubes from graphene, magnetic fields were introduced by Ostrikov et al. [77,78]. Figure 12 shows the schematic diagram of the arc plasma method in the magnetic field. It can be observed that the carbon nanotubes and graphene are deposited on different area of a magnet. The effects of the experimental parameters are also discussed. The effect of pressure changing from 400 Torr to 1000 Torr was reported in 2010 and the results showed that with the increase of experimental pressure, graphene with higher quality would be achieved [79]. In addition, the effect of buffer gas, which acts as coolant, has already been studied [80]. As mentioned before, the control mechanism should be further investigated to meet the needs of industrialization.

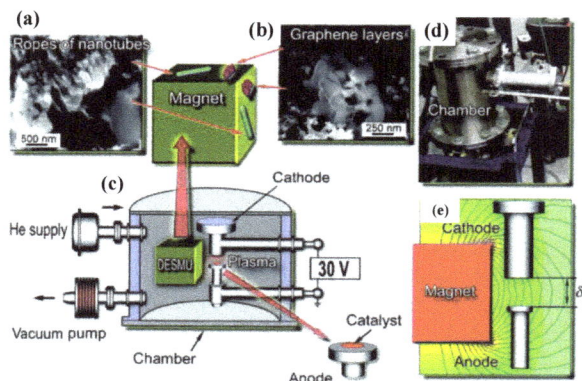

Figure 12. Schematic diagram of arc plasma method in the magnetic field. (**a**) Carbon nanotubes; (**b**) Graphene; (**c**) Experimental structure; (**d**) Real equipment; (**e**) Local image. (Reproduced from [77] with permission; Copyright 2010 The Royal Society of Chemistry.)

2.3. Combustion Method

The combustion method is regarded as one of the greatest potential technologies for producing carbon materials at a large scale. A homemade set-up was built to synthesize graphene, the mixture gases of oxygen and ethylene were injected into a hydrostatic column with a press-seal of a maximum capacity 16.6 L, and then the spark plug was ignited for combustion. After that, graphene was achieved. Moreover, the different mixture ratios of oxygen and ethylene were investigated, and the ideal ratio was 0.6. When the combustion temperature is below 4000 K, the synthesized product is charcoal instead of graphene [81,82]. A much simpler experimental facility was invented in Tsinghua University, and this set-up consisted of alcohol burner and butane lamp. The alcohol burner was used to provide a protective flame atmosphere, meanwhile the carbon atoms and the needed temperature were produced by butane lamp. The graphene finally deposited on the surface of Ni substrate in a short time [83]. The experimental facility was improved by Tse et al., and the much lower oxygen content graphene was synthesized [84].

Another combustion method has been studied by Xiao et al. [85], who put magnesium (Mg) into a carbon dioxide (CO_2) atmosphere. After combusting, the black product was analyzed by TEM and

Raman spectrum and the material was proved to be graphene with fewer than 10 layers, as shown in Figure 13. The mechanism is discussed and the reactive progress is shown as Equation (1). Graphene can also be achieved by combustion of calcium (Ca) in the presence of CO_2 and the mechanism is shown in Equation (2) [86].

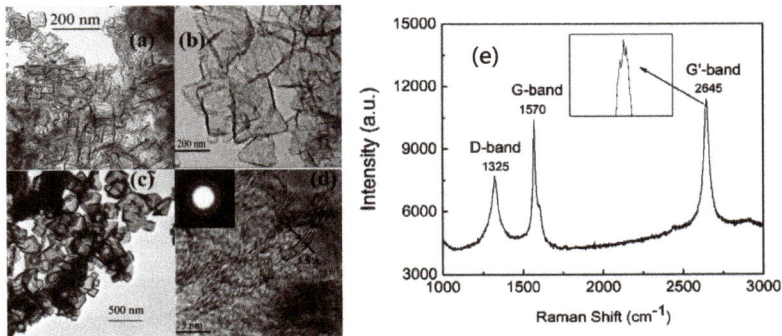

$$2Mg + CO_2 = 2MgO + C \tag{1}$$

$$2Ca + CO_2 = 2CaO + C \tag{2}$$

Figure 13. Characterized images of graphene via combusting Mg in the CO_2 atmosphere. (a–d) TEM images under different scales; (e) Raman spectrum. (Reproduced from [85] with permission; Copyright 2011 The Royal Society of Chemistry.)

How to choose these three methods is also a question, thus a comparison of experimental temperature, energy cost, throughput and electrical properties of different synthesis methods is given in Table 4. This table is beneficial for further consideration of the cost and the return on investment in different methods.

Table 4. A comparison of each growth technique in comprehensive condition.

Method		Temperature	Energy Cost	Throughput	Electrical Property
Epitaxial growth		Medium	High	Low	High
CVD	CCVD	Medium	High	Low	High
	PECVD	Low	Medium	Low	High
	Arc discharge	High	Low	High	Low
Combustion method		High	Low	–	Low

3. Conclusions

The progress in synthesis of high-quality and large-scale graphene is worthy of praise, especially in the field of fundamental research. However, the challenge still exists in the real world. The structural controllability and membrane homogeneity are big puzzles that researchers are faced with. In this paper, the mechanism of the thermal growth method has been discussed in detail and three different thermal growth methods have been presented. These methods are all based on splitting decomposition of molecules and recrystallization of atoms. Epitaxial growth of graphene, which synthesize graphene in relative high quality, is widely used in electronic products due to the possibility of graphene being directly grown on SiC. Chemical vapor deposition possesses the unique advantage of producing large-scale graphene, and becomes a research priority. The combustion method has the lowest requirement for experimental facility and an open experimental environment, which catches investigators' attentions. However, the quality of graphene made by this method is very poor.

Nowadays, new 2D materials, such as h-BN and MoS2, have been used to act as a substrate to synthesize graphene, and these products have much better performance. That is to say, proper optimizing experimental parameters, such as deposition time and temperature, can improve the quality of graphene. The introduction of plasma and roll-to-roll technology has proved that new technology may promote the development of graphene synthesis. Researchers should pay attention not only to the existing methods, but also to the latest technologies. Novel combinations of existing technologies are needed. In order to reduce cost and improve the quality of transferred graphene, it is better to synthesize graphene on target substrates, or a new and undamaged transfer technology should be presented. More efforts should also be made on the reduction of energy consumption and the simplification of equipment.

Acknowledgments: This research is supported by the Beijing Natural Science Foundation (No. 3162024), Tribology Science Foundation of State Key laboratory of Tribology (No. SKLTKF14A08), and Science Foundation of China University of Petroleum, Beijing (No. 2462017BJB06, C201602).

Conflicts of Interest: The authors declare no conflict of interest.

References

1. Peierls, R.E. Bemerkungen über umwandlungstemperaturen. *Helv. Phys. Acta* **1934**, *7*, 81–83. (In German)
2. Peierls, R.E. Quelques proprietes typiques des corpses solides. *Ann. I. H. Poincare* **1935**, *5*, 177–222. (In German)
3. Landau, L.D. Zur Theorie der phasenumwandlungen II. *Phys. Z. Sowjetunion* **1937**, *11*, 26–35. (In German)
4. Novoselov, K.S.; Geim, A.K.; Morozov, S.V.; Jiang, D.; Zhang, Y.; Dubonos, S.V.; Grigorieva, I.V.; Firsov, A.A. Electric field effect in atomically thin carbon films. *Science* **2004**, *306*, 666–669. [CrossRef] [PubMed]
5. Kroto, H.W.; Heath, J.R.; O'Brien, S.C.; Curl, R.F.; Smalley, R.E. C60: Buckminsterfullerene. *Nature* **1985**, *318*, 162–163. [CrossRef]
6. Iijima, S. Helical microtubules of graphitic carbon. *Nature* **1991**, *354*, 56–58. [CrossRef]
7. Zhang, Y.; Tan, Y.W.; Stormer, H.L.; Kim, P. Experimental observation of the quantum hall effect and berry's phase in graphene. *Nature* **2005**, *438*, 201–204. [CrossRef] [PubMed]
8. Sweetman, M.J.; May, S.; Mebberson, N.; Pendleton, P.; Vasilev, K.; Plush, S.E.; Hayball, J.D. Activated carbon, carbon nanotubes and graphene: Materials and composites for advanced water purification. *C* **2017**, *3*, 18. [CrossRef]
9. Liu, B.; Xie, J.; Ma, H.; Zhang, X.; Pan, Y.; Lv, J.; Ge, H.; Ren, N.; Su, H.; Xie, X.; et al. From graphite to graphene oxide and graphene oxide quantum dots. *Small* **2017**, *13*. [CrossRef] [PubMed]
10. Buzaglo, M.; Bar, I.P.; Varenik, M.; Shunak, L.; Pevzner, S.; Regev, O. Graphite-to-Graphene: Total conversion. *Adv. Mater.* **2017**, *29*. [CrossRef] [PubMed]
11. Mayorov, A.S.; Gorbachev, R.V.; Morozov, S.V.; Britnell, L.; Jalil, R.; Ponomarenko, L.A.; Blake, P.; Novoselov, K.S.; Watanabe, K.; Taniguchi, T.; et al. Micrometer-scale ballistic transport in encapsulated graphene at room temperature. *Nano Lett.* **2011**, *11*, 2396–2399. [CrossRef] [PubMed]
12. Lee, C.; Wei, X.; Kysar, J.W.; Hone, J. Measurement of the elastic properties and intrinsic strength of monolayer graphene. *Science* **2008**, *321*, 385–388. [CrossRef] [PubMed]
13. Balandin, A.A.; Ghosh, S.; Bao, W.; Calizo, I.; Teweldebrhan, D.; Miao, F.; Lau, C.N. Superior thermal conductivity of single-layer graphene. *Nano Lett.* **2008**, *8*, 902–907. [CrossRef] [PubMed]
14. Nair, R.R.; Blake, P.; Grigorenko, A.N.; Novoselov, K.S.; Booth, T.J.; Stauber, T.; Peres, N.M.R.; Geim, A.K. Fine structure constant defines visual tranparency of graphene. *Science* **2008**, *320*, 1308. [CrossRef] [PubMed]
15. Yi, M.; Shen, Z. A review on mechanical exfoliation for the scalable production of graphene. *J. Mater. Chem. A* **2015**, *3*, 11700–11715. [CrossRef]
16. Ponomarenko, L.A.; Schedin, F.; Katsnelson, M.I.; Yang, R.; Hill, E.W.; Novoselov, K.S.; Geim, A.K. Chaotic Dirac billiard in graphene quantum dots. *Science* **2008**, *320*, 356–358. [CrossRef] [PubMed]
17. Lin, Y.M.; Dimitrakopoulos, C.; Jenkins, K.A.; Farmer, D.B.; Chiu, H.Y.; Grill, A.; Avouris, P. 100-GHz transistors from wafer-scale epitaxial graphene. *Science* **2010**, *327*, 662. [CrossRef] [PubMed]
18. Badami, D.V. Graphitization of α-silicon carbide. *Nature* **1962**, *193*, 569–570. [CrossRef]

19. Hass, J.; Millán-Otoya, J.E.; First, P.N.; Conrad, E.H. Interface structure of epitaxial graphene grown on 4H-SiC (0001). *Phys. Rev. B* **2008**, *78*, 205424. [CrossRef]

20. Jernigan, G.G.; VanMil, B.L.; Tedesco, J.L.; Tischler, J.G.; Glaser, E.R.; Davidson, A., III; Campbell, P.M.; Gaskill, D.K. Comparison of epitaxial graphene on Si-face and C-face 4H SiC formed by ultrahigh vacuum and RF furnace production. *Nano Lett.* **2009**, *9*, 2605–2609. [CrossRef] [PubMed]

21. Berger, C.; Song, Z.; Li, T.; Li, X.; Ogbazghi, A.Y.; Feng, R.; Dai, Z.; Marchenkov, A.N.; Conrad, E.H.; First, P.N.; et al. Ultrathin epitaxial graphite: 2D electron gas properties and a route toward graphene-based nanoelectronics. *J. Phys. Chem. B* **2004**, *108*, 19912–19916. [CrossRef]

22. Berger, C.; Song, Z.; Li, X.; Wu, X.; Brown, N.; Naud, C.; Mayou, D.; Li, T.; Hass, J.; Marchenkov, A.N.; et al. Electronic confinement and coherence in patterned epitaxial graphene. *Science* **2006**, *312*, 1191–1196. [CrossRef] [PubMed]

23. Kang, C.Y.; Fan, L.L.; Chen, S.; Liu, Z.L.; Xu, P.S.; Zou, C.W. Few-layer graphene growth on 6H-SiC (0001) surface at low temperature via Ni-silicidation reactions. *Appl. Phys. Lett.* **2012**, *100*, 251604. [CrossRef]

24. Tang, J.; Liu, Z.L.; Kang, C.Y.; Yan, W.S.; Xu, P.S.; Pan, H.B.; Wei, S.Q.; Gao, Y.Q.; Xu, X.G. Annealing time dependence of morphology and structure of epitaxial graphene on 6H-SiC(0001) surface. *Acta Phys-Chim. Sin.* **2010**, *26*, 253–258. (In Chinese)

25. Emtsev, K.V.; Bostwick, A.; Horn, K.; Jobst, J.; Kellogg, G.L.; Ley, L.; McChesney, J.L.; Ohta, T.; Reshanov, S.A.; Röhrl, J.; et al. Towards wafer-size graphene layers by atmospheric pressure graphitization of silicon carbide. *Nat. Mater.* **2009**, *8*, 203–207. [CrossRef] [PubMed]

26. Heer, W.A.D.; Berger, C.; Wu, X.; First, P.N.; Conrad, E.H.; Li, X.; Li, T.; Sprinkle, M.; Hass, J.; Sadowski, M.L.; et al. Epitaxial graphene. *Solid State Commun.* **2007**, *143*, 92–100. [CrossRef]

27. Hass, J.; De Heer, W.A.; Conrad, E.H. The growth and morphology of epitaxial multilayer graphene. *J. Phys. Condens. Matter* **2008**, *20*, 323202. [CrossRef]

28. Hicks, J.; Shepperd, K.; Wang, F.; Conrad, E.H. The structure of graphene grown on the SiC surface. *J. Phys. D Appl. Phys.* **2012**, *45*, 154002. [CrossRef]

29. Terai, M.; Hasegawa, N.; Okusawa, M.; Otani, S.; Oshima, C. Electronic states of monolayer micrographite on TiC (111)-faceted and TiC (410) surfaces. *Appl. Surf. Sci.* **1998**, *130*, 876–882. [CrossRef]

30. Itchkawitz, B.S.; Lyman, P.F.; Ownby, G.W.; Zehner, D.M. Monolayer graphite on TaC (111): Electronic band structure. *Surf. Sci.* **1994**, *318*, 395–402. [CrossRef]

31. Yang, W.; Chen, G.; Shi, Z.; Liu, C.C.; Zhang, L.; Xie, G.; Cheng, M.; Wang, D.; Yang, R.; Shi, D.; et al. Epitaxial growth of single-domain graphene on hexagonal boron nitride. *Nat. Mater.* **2013**, *12*, 792–797. [CrossRef] [PubMed]

32. Reina, A.; Jia, X.; Ho, J.; Nezich, D.; Son, H.; Bulovic, V.; Dresselhaus, M.S.; Kong, J. Large area, few-layer graphene films on arbitrary substrates by chemical vapor deposition. *Nano Lett.* **2009**, *9*, 30–35. [CrossRef] [PubMed]

33. Kim, K.S.; Zhao, Y.; Jang, H.; Lee, S.Y.; Kim, J.M.; Kim, K.S.; Ahn, J.H.; Kim, P.; Choi, J.Y.; Hong, B.H. Large-scale pattern growth of graphene films for stretchable transparent electrodes. *Nature* **2009**, *457*, 706–710. [CrossRef] [PubMed]

34. Seo, D.H.; Pineda, S.; Fang, J.; Gozukara, Y.; Yick, S.; Bendavid, A.; Lam, S.K.H.; Murdock, A.T.; Murphy, A.B.; Han, Z.J.; et al. Single-step ambient-air synthesis of graphene from renewable precursors as electrochemical genosensor. *Nat. Commun.* **2017**, *8*, 14217. [CrossRef] [PubMed]

35. Li, X.; Cai, W.; An, J.; Kim, S.; Nah, J.; Yang, D.; Piner, R.; Velamakanni, A.; Jung, I.; Tutuc, E.; et al. Large-area synthesis of high-quality and uniform graphene films on copper foils. *Science* **2009**, *324*, 1312–1314. [CrossRef] [PubMed]

36. Bae, S.; Kim, H.; Lee, Y.; Xu, X.; Park, J.S.; Zheng, Y.; Balakrishnan, J.; Lei, T.; Kim, H.R.; Song, Y., II; et al. Roll-to-roll production of 30-inch graphene films for transparent electrodes. *Nat. Nanotechnol.* **2010**, *5*, 574–578. [CrossRef] [PubMed]

37. Xu, J.; Fu, C.; Sun, H.; Meng, L.; Xia, Y.; Zhang, C.; Yi, X.; Yang, W.; Guo, P.; Wang, C.; et al. Large-area, high-quality monolayer graphene from polystyrene at atmospheric pressure. *Nanotechnology* **2017**, *28*, 155605. [CrossRef] [PubMed]

38. Malesevic, A.; Vitchev, R.; Schouteden, K.; Volodin, A.; Zhang, L.; Van Tendeloo, G.; Vanhulsel, A.; Van Haesendonck, C. Synthesis of few-layer graphene via microwave plasma-enhanced chemical vapor deposition. *Nanotechnology* **2008**, *19*, 305604. [CrossRef] [PubMed]

39. Boyd, D.A.; Lin, W.H.; Hsu, C.C.; Teague, M.L.; Chen, C.C.; Lo, Y.Y.; Chan, W.Y.; Su, W.B.; Cheng, T.C.; Chang, C.S.; et al. Single-step deposition of high-mobility graphene at reduced temperatures. *Nat. Commun.* **2015**, *6*, 6620. [CrossRef] [PubMed]

40. Dato, A.; Radmilovic, V.; Lee, Z.; Phillips, J.; Frenklach, M. Substrate-free gas-phase synthesis of graphene sheets. *Nano Lett.* **2008**, *8*, 2012–2016. [CrossRef] [PubMed]

41. Subrahmanyam, K.S.; Panchakarla, L.S.; Govindaraj, A.; Rao, C.N.R. Simple method of preparing graphene flakes by an arc-discharge method. *J. Phys. Chem. C* **2009**, *113*, 4257–4259. [CrossRef]

42. Son, M.; Lim, H.; Hong, M.; Choi, H.C. Direct growth of graphene pad on exfoliated hexagonal boron nitride surface. *Nanoscale* **2011**, *3*, 3089–3093. [CrossRef] [PubMed]

43. Tang, S.; Ding, G.; Xie, X.; Chen, J.; Wang, C.; Ding, X.; Huang, F.; Lu, W.; Jiang, M. Nucleation and growth of single crystal graphene on hexagonal boron nitride. *Carbon* **2012**, *50*, 329–331. [CrossRef]

44. Liu, Z.; Song, L.; Zhao, S.; Huang, J.; Ma, L.; Zhang, J.; Lou, J.; Ajayan, P.M. Direct growth of graphene/hexagonal boron nitride stacked layers. *Nano Lett.* **2011**, *11*, 2032–2037. [CrossRef] [PubMed]

45. Kwieciński, W.; Sotthewes, K.; Poelsema, B.; Zandvliet, H.J.; Bampoulis, P. Chemical vapor deposition growth of bilayer graphene in between molybdenum disulfide sheets. *J. Colloid Interface Sci.* **2017**, *505*, 776–782. [CrossRef] [PubMed]

46. Fu, W.; Du, F.H.; Su, J.; Li, X.H.; Wei, X.; Ye, T.N.; Wang, K.X.; Chen, J.S. In situ catalytic growth of large-area multilayered graphene/MoS$_2$ heterostructures. *Sci. Rep.* **2014**, *4*, 4673. [CrossRef] [PubMed]

47. Oshima, C.; Nagashima, A. Ultra-thin epitaxial films of graphite and hexagonal boron nitride on solid surfaces. *J. Phys. Condens. Matter* **1997**, *9*, 1. [CrossRef]

48. Kong, J.; Cassell, A.M.; Dai, H. Chemical vapor deposition of methane for single-walled carbon nanotubes. *Chem. Phys. Lett.* **1998**, *292*, 567–574. [CrossRef]

49. Reina, A.; Thiele, S.; Jia, X.; Bhaviripudi, S.; Dresselhaus, M.S.; Schaefer, J.A.; Kong, J. Growth of large-area single-and bi-layer graphene by controlled carbon precipitation on polycrystalline Ni surfaces. *Nano Res.* **2009**, *2*, 509–516. [CrossRef]

50. Zhang, Y.; Gomez, L.; Ishikawa, F.N.; Madaria, A.; Ryu, K.; Wang, C.; Badmaev, A.; Zhou, C. Comparison of graphene growth on single-crystalline and polycrystalline Ni by chemical vapor deposition. *J. Phys. Chem. Lett.* **2010**, *1*, 3101–3107. [CrossRef]

51. Chae, S.J.; Güneş, F.; Kim, K.K.; Kim, E.S.; Han, G.H.; Kim, S.M.; Shin, H.J.; Yoon, S.M.; Choi, J.Y.; Park, M.H.; et al. Synthesis of large-area graphene layers on poly-nickel substrate by chemical vapor deposition: Wrinkle formation. *Adv. Mater.* **2009**, *21*, 2328–2333. [CrossRef]

52. Liu, J.; Li, P.; Chen, Y.; Wang, Z.; He, J.; Tian, H.; Qi, F.; Zheng, B.; Zhou, J.; Lin, W.; et al. Large-area synthesis of high-quality and uniform monolayer graphene without unexpected bilayer regions. *J. Alloys Compd.* **2014**, *615*, 415–418. [CrossRef]

53. Vlassiouk, I.; Smirnov, S.; Regmi, M.; Surwade, S.P.; Srivastava, N.; Feenstra, R.; Eres, G.; Parish, C.; Lavrik, N.; Datskos, P.; et al. Graphene nucleation density on copper: Fundamental role of background pressure. *J. Phys. Chem. C* **2013**, *117*, 18919–18926. [CrossRef]

54. Liu, L.; Zhou, H.; Cheng, R.; Chen, Y.; Lin, Y.C.; Qu, Y.; Bai, J.; Ivanov, I.A.; Liu, G.; Huang, Y.; et al. A systematic study of atmospheric pressure chemical vapor deposition growth of large-area monolayer graphene. *J. Mater. Chem.* **2012**, *22*, 1498–1503. [CrossRef] [PubMed]

55. Li, X.; Cai, W.; Colombo, L.; Ruoff, R.S. Evolution of graphene growth on Ni and Cu by carbon isotope labeling. *Nano Lett.* **2009**, *9*, 4268–4272. [CrossRef] [PubMed]

56. Whiteway, E.; Yang, W.; Yu, V.; Hilke, M. Time evolution of the growth of single graphene crystals and high resolution isotope labeling. *Carbon* **2017**, *111*, 173–181. [CrossRef]

57. Chen, C.; Avila, J.; Asensio, M.C. Chemical and electronic structure imaging of graphene on Cu: A NanoARPES study. *J. Phys. Condens. Matter* **2017**, *29*, 183001. [CrossRef] [PubMed]

58. Yang, F.; Liu, Y.; Wu, W.; Chen, W.; Gao, L.; Sun, J. A facile method to observe graphene growth on copper foil. *Nanotechnology* **2012**, *23*, 475705. [CrossRef] [PubMed]

59. Vinogradov, N.A.; Schulte, K.; Ng, M.L.; Mikkelsen, A.; Lundgren, E.; Martensson, N.; Preobrajenski, A.B. Impact of atomic oxygen on the structure of graphene formed on Ir (111) and Pt (111). *J. Phys. Chem. C* **2011**, *115*, 9568–9577. [CrossRef]

60. Coraux, J.; Engler, M.; Busse, C.; Wall, D.; Buckanie, N.; Zu Heringdorf, F.J.M.; Van Gastel, R.; Poelsema, B.; Michely, T. Growth of graphene on Ir (111). *New J. Phys.* **2009**, *11*, 023006. [CrossRef]

61. Lukosius, M.; Dabrowski, J.; Kitzmann, J.; Fursenko, O.; Akhtar, F.; Lisker, M.; Lippert, G.; Schulze, S.; Yamamoto, Y.; Schubert, M.A.; et al. Metal-free CVD graphene synthesis on 200 mm Ge/Si (001) substrates. *ACS Appl. Mater. Interfaces* **2016**, *8*, 33786–33793. [CrossRef] [PubMed]

62. Scaparro, A.M.; Miseikis, V.; Coletti, C.; Notargiacomo, A.; Pea, M.; De Seta, M.; Di Gaspare, L. Investigating the CVD Synthesis of Graphene on Ge (100): Toward layer-by-layer growth. *ACS Appl. Mater. Interfaces* **2016**, *8*, 33083–33090. [CrossRef] [PubMed]

63. Wang, X.; You, H.; Liu, F.; Li, M.; Wan, L.; Li, S.; Li, Q.; Xu, Y.; Tian, R.; Yu, Z.; et al. Large-scale synthesis of few-layered graphene using CVD. *Chem. Vapor Depos.* **2009**, *15*, 53–56. [CrossRef]

64. Ryu, J.; Kim, Y.; Won, D.; Kim, N.; Park, J.S.; Lee, E.K.; Cho, D.; Cho, S.P.; Kim, S.J.; Ryu, G.H.; et al. Fast synthesis of high-performance graphene films by hydrogen-free rapid thermal chemical vapor deposition. *ACS Nano* **2014**, *8*, 950–956. [CrossRef] [PubMed]

65. Chen, Y.Z.; Medina, H.; Tsai, H.W.; Wang, Y.C.; Yen, Y.T.; Manikandan, A.; Chueh, Y.L. Low temperature growth of graphene on glass by carbon-enclosed chemical vapor deposition process and its application as transparent electrode. *Chem. Mater.* **2015**, *27*, 1646–1655. [CrossRef]

66. Chan, S.H.; Chen, S.H.; Lin, W.T.; Li, M.C.; Lin, Y.C.; Kuo, C.C. Low-temperature synthesis of graphene on Cu using plasma-assisted thermal chemical vapor deposition. *Nanoscale Res. Lett.* **2013**, *8*, 285. [CrossRef] [PubMed]

67. Mueller, N.S.; Morfa, A.J.; Abou-Ras, D.; Oddone, V.; Ciuk, T.; Giersig, M. Growing graphene on polycrystalline copper foils by ultra-high vacuum chemical vapor deposition. *Carbon* **2014**, *78*, 347–355. [CrossRef]

68. Jang, J.; Son, M.; Chung, S.; Kim, K.; Cho, C.; Lee, B.H.; Ham, M.H. Low-temperature-grown continuous graphene films from benzene by chemical vapor deposition at ambient pressure. *Sci. Rep.* **2015**, *5*, 17955. [CrossRef] [PubMed]

69. John, P.I. *Plasma Sciences and the Creation of Wealth*; Tata McGraw-Hill Education: New York, NY, USA, 2005.

70. Qi, J.; Zhang, L.; Cao, J.; Zheng, W.; Wang, X.; Feng, J. Synthesis of graphene on a Ni film by radio-frequency plasma-enhanced chemical vapor deposition. *Chin. Sci. Bull.* **2012**, *57*, 3040–3044. [CrossRef]

71. Karmakar, S.; Kulkarni, N.V.; Nawale, A.B.; Lalla, N.P.; Mishra, R.; Sathe, V.G.; Bhoraskar, S.V.; Das, A.K. A novel approach towards selective bulk synthesis of few-layer graphenes in an electric arc. *J. Phys. D Appl. Phys.* **2009**, *42*, 115201. [CrossRef]

72. Tatarova, E.; Dias, A.; Henriques, J.; do Rego, A.B.; Ferraria, A.M.; Abrashev, M.V.; Luhrs, C.C.; Phillips, J.; Dias, F.M.; Ferreira, C.M. Microwave plasmas applied for the synthesis of free standing graphene sheets. *J. Phys. D Appl. Phys.* **2014**, *47*, 38550. [CrossRef]

73. Park, H.J.; Ahn, B.W.; Kim, T.Y.; Lee, J.W.; Jung, Y.H.; Choi, Y.S.; Song, Y., II; Suh, S.J. Direct synthesis of multi-layer graphene film on various substrates by microwave plasma at low temperature. *Thin Solid Films* **2015**, *587*, 8–13. [CrossRef]

74. Yamada, T.; Ishihara, M.; Kim, J.; Hasegawa, M.; Iijima, S. A roll-to-roll microwave plasma chemical vapor deposition process for the production of 294 mm width graphene films at low temperature. *Carbon* **2012**, *50*, 2615–2619. [CrossRef]

75. Phillips, J. Plasma Generation of Supported Metal Catalysts. U.S. Patent 5,989,648, 23 November 1999.

76. Dato, A.M. *Substrate-Free Gas-Phase Synthesis of Graphene*; University of California: Berkeley, CA, USA, 2009.

77. Volotskova, O.; Levchenko, I.; Shashurin, A.; Raitses, Y.; Ostrikov, K.; Keidar, M. Single-step synthesis and magnetic separation of graphene and carbon nanotubes in arc discharge plasmas. *Nanoscale* **2010**, *2*, 2281–2285. [CrossRef] [PubMed]

78. Levchenko, I.; Volotskova, O.; Shashurin, A.; Raitses, Y.; Ostrikov, K.; Keidar, M. The large-scale production of graphene flakes using magnetically-enhanced arc discharge between carbon electrodes. *Carbon* **2010**, *48*, 4570–4574. [CrossRef]

79. Wang, Z.; Li, N.; Shi, Z.; Gu, Z. Low-cost and large-scale synthesis of graphene nanosheets by arc discharge in air. *Nanotechnology* **2010**, *21*, 175602. [CrossRef] [PubMed]

80. Shen, B.; Ding, J.; Yan, X.; Feng, W.; Li, J.; Xue, Q. Influence of different buffer gases on synthesis of few-layered graphene by arc discharge method. *Appl. Surf. Sci.* **2012**, *258*, 4523–4531. [CrossRef]

81. Sorensen, C.; Nepal, A.; Singh, G.P. Process for High-Yield Production of Graphene via Detonation of Carbon-Containing Material. U.S. Patent 9,440,857, 13 September 2016.

82. Nepal, A.; Singh, G.P.; Flanders, B.N.; Sorensen, C.M. One-step synthesis of graphene via catalyst-free gas-phase hydrocarbon detonation. *Nanotechnology* **2013**, *24*, 245602. [CrossRef] [PubMed]

83. Li, Z.; Zhu, H.; Xie, D.; Wang, K.; Cao, A.; Wei, J.; Li, X.; Fan, L.; Wu, D. Flame synthesis of few-layered graphene/graphite films. *Chem. Commun.* **2011**, *47*, 3520–3522. [CrossRef] [PubMed]

84. Memon, N.K.; Stephen, D.T.; Al-Sharab, J.F.; Yamaguchi, H.; Goncalves, A.M.B.; Kear, B.H.; Jaluria, Y.; Andrei, E.Y.; Chhowalla, M. Flame synthesis of graphene films in open environments. *Carbon* **2011**, *49*, 5064–5070. [CrossRef]

85. Chakrabarti, A.; Lu, J.; Skrabutenas, J.C.; Xu, T.; Xiao, Z.; Maguire, J.A.; Hosmane, N.S. Conversion of carbon dioxide to few-layer graphene. *J. Mater. Chem.* **2011**, *21*, 9491–9493. [CrossRef]

86. Zhang, J.; Tian, T.; Chen, Y.; Niu, Y.; Tang, J.; Qin, L.C. Synthesis of graphene from dry ice in flames and its application in supercapacitors. *Chem. Phys. Lett.* **2014**, *591*, 78–81. [CrossRef]

coatings

MDPI

Review

Recent Advances in Graphene-Based Free-Standing Films for Thermal Management: Synthesis, Properties, and Applications

Feng Gong [1,*], Hao Li [1,†], Wenbin Wang [1,†], Dawei Xia [1], Qiming Liu [1], Dimitrios V. Papavassiliou [2,*] and Ziqiang Xu [1]

[1] School of Materials and Energy, University of Electronic Science and Technology of China, Chengdu 611731, China; 15520768039m@sina.cn (H.L.); bingo_uestc@163.com (W.W.); davidxia97@163.com (D.X.); qimingliuwhr@gmail.com (Q.L.); nanterxu@uestc.edu.cn (Z.X.)
[2] School of Chemical, Biological, and Materials Engineering, University of Oklahoma, Norman, OK 73019, USA
* Correspondence: gongfeng@uestc.edu.cn (F.G.); dvpapava@ou.edu (D.V.P.)
† These authors contributed equally to this work.

Received: 20 January 2018; Accepted: 4 February 2018; Published: 7 February 2018

Abstract: Thermal management in microelectronic devices has become a crucial issue as the devices are more and more integrated into micro-devices. Recently, free-standing graphene films (GFs) with outstanding thermal conductivity, superb mechanical strength, and low bulk density, have been regarded as promising materials for heat dissipation and for use as thermal interfacial materials in microelectronic devices. Recent studies on free-standing GFs obtained via various approaches are reviewed here. Special attention is paid to their synthesis method, thermal conductivity, and potential applications. In addition, the most important factors that affect the thermal conductivity are outlined and discussed. The scope is to provide a clear overview that researchers can adopt when fabricating GFs with improved thermal conductivity and a large area for industrial applications.

Keywords: graphene; free-standing films; thermal conductivity; thermal management

1. Introduction

Small yet powerful micro/nano-electronic devices used in electronic, portable, and automotive products generate a large amount of heat during operation, which may cause damage or failure [1–3]. There is a critical demand for methods and materials that can be used to dissipate the generated heat from microelectronic devices. As electronic devices become more and more miniaturized, this need is becoming more and more urgent. The current materials used for thermal dissipation in microelectronic devices are copper or aluminum alloys, which attain a thermal conductivity of 401 W/m·K and 121 W/m·K, respectively [4]. However, the high cost, poor processing techniques, and high density of these metals can hinder their broad application in micro- or nano-scale electronic devices.

Graphene, in monolayer or few-layer forms, has been demonstrated to reach ultrahigh levels of thermal conductivity (up to ~5000 W/m·K) by both experiments and simulations [5–9]. Nevertheless, there are few reliable approaches to handle such atomically thin materials, which hampers their device-related applications [10]. Many researchers have worked to incorporate graphene into polymer matrices for nanocomposites with advanced functionality [11–16]. These nanocomposites achieve low density and high flexibility, as well as enhanced thermal conductivity, and are processing technologies that can be implemented fairly easily. Yet, limited by the undesirable thermal conductivity of polymer matrices (~0.1 W/m·K) and the high interfacial thermal resistance between graphene and polymer, the thermal conductivity of graphene nanocomposites (~1 W/m·K) is still much lower than expected [17–21]. This low thermal conductivity of graphene nanocomposites makes it difficult to satisfy the requirement of efficient heat dissipation in micro/nano-electronic devices.

Recently, free-standing films obtained from graphene or graphene oxide (GO) have attracted loads of interest in the thermal management field due to their high thermal conductivity, superior electrical conductivity, and excellent mechanical properties [22–24]. Free-standing graphene films (GFs) could inherit the ultrahigh thermal conductivity of graphene, achieving a thermal conductivity above 3000 W/m·K [25]. Compared to the metals commonly used for heat dissipation, GFs possess a much lower density (<1 g/cm^3), superb flexibility as well as lower cost, demonstrating the potential for effective heat dissipation and use as thermal interfacial materials. In the past few years, numerous studies have been carried out to fabricate GFs with high thermal conductivity. However, it is noted that through different synthesis methods the GFs achieved substantially dissimilar thermal conductivity, varying from 30 to 3300 [25,26]. Thus, it is necessary to review recent advances in free-standing GFs to inform the design of GFs with advanced functionality.

In this review, we summarize the state of the art with regard to free-standing GFs in recent years. We focus on fabrication approaches as well as the thermal conductivity of the as-prepared GFs. The factors most influencing the thermal conductivity of GFs are identified and the impact of these factors is discussed. Several typical applications of the GFs for thermal management are also epitomized and discussed. The goal is to provide a concise overview of designing free-standing GFs with improved thermal conductivity.

2. Synthesis of Graphene-Based Films

There are several approaches to obtain graphene films with thickness in microns [27–29]. Similar to the fabrication of graphene, synthesis of graphene films follows two main procedures: "top-down" and "bottom-up". The top-down processes attain graphene films from graphite, including vacuum filtration [26,27,30–36], direct evaporation [37–39], and diverse coating techniques [28,40]. The bottom-up processes are based on the synthesis of graphene films from gaseous carbon sources, such as chemical vapor deposition [23].

2.1. Vacuum Filtration Method

Vacuum filtration is a convenient and widely-used top-down approach to assemble GFs. The pore size of filter films is controlled to make sure that the solvent molecules can permeate the filter paper easily, while graphene sheets remain on the surface of filter film. The typical process of the vacuum filtration strategy is illustrated in Figure 1. Graphite is adopted as the source material. Graphene oxide (GO) is first obtained from graphite powders via the well-established Hummers' method or modified Hummers' method [41]. Then GO solutions with various GO concentrations are prepared for the filtration. The thickness of the GO films can be tuned by modulating the GO concentration. The wet GO films can be fabricated after vacuum filtration for several minutes. Then, the GO film can be easily peeled off from the filter film after being dried at 80 °C for several minutes. In this stage, GO films show very low thermal conductivity due to the intrinsic low thermal conductivity of the GO. Thus, the GO films have to be reduced to GFs via high-temperature annealing. The annealing temperatures are as high as 3000 °C [28]. Such high temperatures can effectively graphitize the GO to graphene, leading to improved thermal conductivity of the final GFs. However, the use of such a high temperature for graphitization may hinder the wide application of GFs.

Not only the GO solution, but also the graphene solution can be directly applied in vacuum filtration for the GFs. For example, Teng et al. [34] applied a facile ball-milling approach to obtain large-volume, high-concentration, and plane-defect-free graphene dispersion in *N*-methyl-2-pyrrolidone (NMP) from graphite. After the filtration, annealing at 2850 °C for 2 h and a compression at 30 MPa, the obtained GFs exhibited superb electrical conductivity of 2.23 × 10^5 S/cm, and high thermal conductivity of 1529 W/m·K [34].

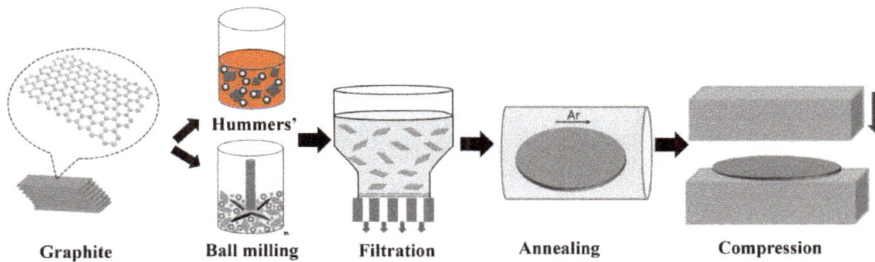

Figure 1. Schematic plot of the process of vacuum filtration for graphene films. Graphite can be oxidized to GO using Hummers' method, or exfoliated in ball milling. Reproduced and modified from [34] with permission; Copyright 2017 Wiley Online Library.

2.2. Direct Evaporation Method

The direct evaporation method is another simple approach to GF fabrication. In this method, the GO solution is poured into a container with a flat bottom, followed by continuous evaporation by heating the GO solution to a proper temperature. After several hours, the solvent (i.e., water) is evaporated and a thin film is left on the surface of the container. For instance, Shen et al. [37] used the direct evaporation method to attain the GFs. The GO solution was obtained by dissolving GO in water and the GO solution was poured into Teflon dishes for evaporation. Then, the Teflon dishes were heated to 50–60 °C for 6–10 h to evaporate the water and form GO films. The GO film was annealed at 2000 °C under argon flow for one hour after peeling off from the dishes. The dark, freestanding, and flexible GFs were obtained after the high-temperature annealing. The GFs have a superb in-plane thermal conductivity of about 1100 W/m·K. Similarly, Chen et al. applied the evaporation method to form GFs at the liquid/air interface [39]. The GO suspension was sonicated to exfoliate GO sheets and centrifuged to remove impurities. The obtained GO hydrosol was heated to 80 °C for different times to get films of different thicknesses. The condensed thin film formed very rapidly at the liquid/air interface. After drying at 80 °C for 8 h, smooth and free-standing GFs were obtained.

The GO solution or graphene solution with high concentration can be directly coated onto a substrate for the evaporation step, rather than poured in containers. Recently, Peng et al. [28] scraped a GO suspension (10–20 mg·mL^{-1}) with a thickness of 0.5–5 mm on copper foil and kept evaporating the water at room temperature for 24 h. The as-prepared GO film was carbonized at 1300 °C for 2 h and graphitized at a high temperature of 3000 °C for 1 h under argon flow. After cooling slowly to room temperature, the GF was compressed under a pressure of 50 MPa for 15 min, 100 MPa for 30 min, and 300 MPa for 1 h to form a dense GF. In this method, GFs with a large surface area can be easily obtained by utilizing copper foil. The copper foil can also be recycled for low-cost, large-scale fabrication of GFs [42].

It is noted that in vacuum filtration and direct evaporation, thermal annealing is not the only way to reduce a GO film to graphene films. Chemical reduction is also a broadly-used approach to reduce GO to rGO. The reducing agents can be either organic or inorganic chemicals, such as ascorbic acid and HI acid [43–46]. Yang et al. applied HI acid to reduce the GO/cellulose composite film from vacuum filtration [47]. The obtained composite film exhibited an in-plane thermal conductivity of 7.3 W/m·K, and a strong anisotropy in thermal conductivity. Jin et al. compared the effect of different reduction methods on the thermal conductivity of the GFs [48]. They used HI acid, thermal reduction at 600 °C and the combination of HI acid and thermal conduction to reduce the GO films to the GFs. It is demonstrated that thermal reduction is the best way to achieve GFs with higher thermal conductivity, compared with other reduction methods.

2.3. Other Assembling Methods

Besides the aforementioned vacuum filtration and direct evaporation methods, there are also some other top-down methods to fabricate GFs, such as spray coating [24,49,50], spin casting [51,52], and roll-to-roll manufacturing [53]. For instance, Xin et al. [40] adopted a novel approach integrating electron-spray deposition (ESD) and a roll-to-roll device to manufacture the free-standing GFs with large area; the process is illustrated in Figure 2. In the ESD process (Figure 2a), tiny droplets were generated by the repulsion force between electrical charges and the droplets, while the size of the droplets could be controlled by adjusting the flow speed and electric field. The electric field between nozzle and substrate allowed the droplets to distribute uniformly on the surface of the substrate. Simultaneously, the heating plate could evaporate the solvent, leaving the GFs on the surface of the substrate. A roll-to-roll device was employed to obtain GFs with a large area, shown in Figure 2b. This combined facility opens up the possibility of manufacturing GFs on a large scale, which may be adopted in the industrial manufacturing of GFs.

In summary, for the top-down strategies for GFs, we can conclude that all of them obtain GFs from GO or graphene nanosheets via different assembling techniques. To achieve high thermal conductivity of the GFs, high-temperature thermal annealing or chemical reduction is often conducted.

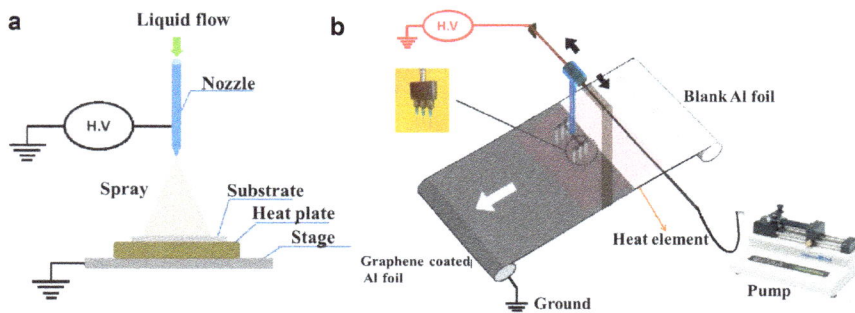

Figure 2. Schematic plot of (**a**) electron-spray deposition and (**b**) roll-to-roll device with a pump. Reproduced from [40] with permission; Copyright 2014 Wiley Online Library.

2.4. Chemical Vapor Deposition

Chemical vapor deposition (CVD) is the typical bottom-up strategy to fabricate a high-quality graphene film with controllable thickness. Ma et al. developed a segregation-adsorption CVD (SACVD) to grow a well-stitched high-quality monolayer graphene film with a tunable uniform grain size from 200 nm to 1 μm on a Pt substrate [23]. They found that the thermal and electrical conductivity of the GF could be tuned by modulating the grain size of graphene. The high-quality graphene film from SACVD with tunable thermal and electrical conductivity can be directly used in electronic, optoelectronic, and thermoelectric applications. However, the transfer of GFs from the Pt substrates may be difficult, limiting their broad application. Thus, the non-substrate CVD approach is preferred to effectively fabricate GFs [54]. Hu et al. combined CVD and spray coating to synthesize CNT/graphene composite films [54]. Spinnable CNT arrays were continuously synthesized and collected by a rotating mandrel from a CVD furnace. During the collection of CNT arrays, the GO aqueous solution was sprayed onto the winding mandrel, as illustrated in Figure 3. After thermal annealing at 2800 °C, the composite films achieved a high thermal conductivity of 1056 W/m·K and a superior mechanical strength of ~1 GPa. A combination of CVD and other simple techniques can be used for the large-scale fabrication of high-quality GFs.

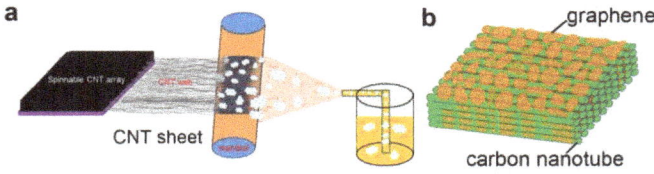

Figure 3. Schematic plot of CVD and the spray-coating process (**a**), and the obtained free-standing CNT array/GO composite films (**b**). Reprinted from [54] with permission; Copyright 2017 Elsevier.

3. Thermal Transport Properties of Free-Standing Graphene Films

3.1. Macro- and Micro-Scale Morphologies of Free-Standing Graphene Films

The GFs fabricated by different methods normally exhibited different macro-scale morphology, but similar micro-scale morphology due to the involved high-temperature annealing. As shown in Figure 4a, after vacuum filtration the GO films displayed a dark brown color and had a rounded shape due to the filtration membranes used in the process [36]. After reduction using HI acid, the rGO films had a shiny metallic color (Figure 4b) and excellent flexibility (Figure 4c). Scanning electron microscope (SEM) images showed that the thickness of the GO films after vacuum filtration was around 10 μm (Figure 4d) and the thickness decreased to ~7.5 μm after the HI reduction (Figure 4e). The layered structures can be clearly observed in the SEM cross section images. Different from the GFs obtained via the chemical reduction, the GFs experiencing high-temperature annealing were normally gray and had wrinkles on the surface [48]. The thermal annealing also caused air pores in the GFs (Figure 5), decreasing the thermal conductivity of the GFs. Therefore, high-pressure compression is always conducted to remove air pores from the GFs. Compared to the GFs obtained by the vacuum filtration method, the GFs from other top-down strategies can have a larger area and a different shape [22]. All these GFs obtained from the top-down strategies show excellent flexibility, as seen in Figure 6, where the GFs can be folded into complex structures [22].

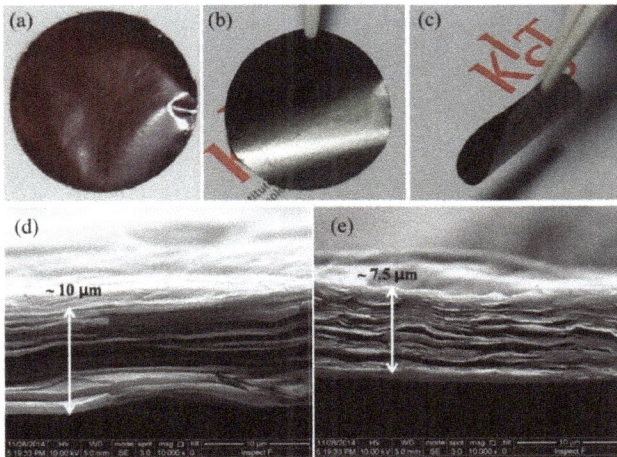

Figure 4. Macro-scale morphology of (**a**) the free-standing dark-brown GO film; (**b**,**c**) are HI reduced shiny metallic and flexible rGO films; (**d**,**e**) are cross-sectional SEM images of as-prepared GO and rGO films, respectively. Reprinted from [36] with permission; Copyright 2015 Elsevier.

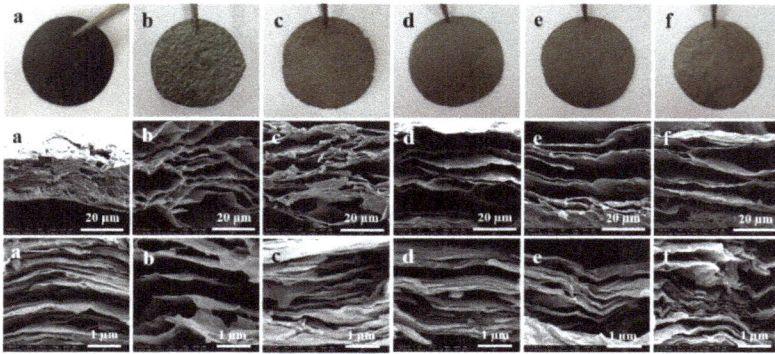

Figure 5. Visual (top), low-magnification SEM (middle), and high-magnification SEM (bottom) images of (**a**) GO film; (**b**) rGO film with HI recution; (**c**) rGO film with HI reduction + 600 °C reduction; (**d**) rGO film with 600 °C reduction; (**e**) rGO film with 800 °C reduction; and (**f**) rGO film with 1000 °C reduction. Reprinted from [48] with permission; Copyright 2015 Elsevier.

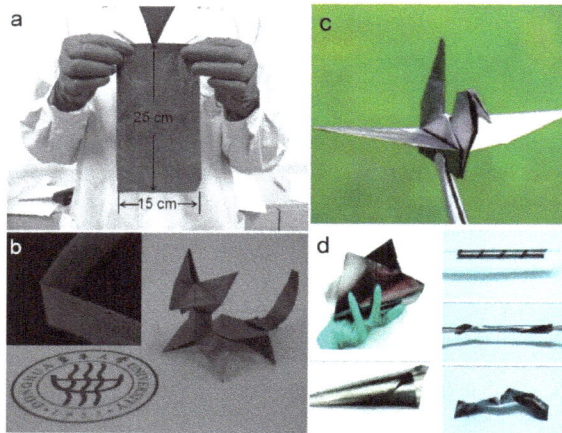

Figure 6. (**a**) Large-area and free-standing GF from direct evaporation; (**b**) sharply folded and twisted craft from the GF; (**c**) a piece of GF was folded into a crane without breakages; and (**d**) the GF in the states of bending, curling, enwinding, twisting and knotting. (**a,b**) Reprinted from [22] with permission, Copyright 2017 Wiley Online Library; (**c,d**) Reprinted from [28] with permission, Copyright 2017 Wiley Online Library.

3.2. Thermal Conductivity of Free-Stranding Graphene Films

Graphene has been demonstrated to have ultrahigh thermal conductivity [46,55,56]. Baladin et al. measured the thermal conductivity of a graphene monolayer and found that it exceeded 3000 W/m·K near room temperature via optothermal Raman measurement [57,58]. A recent study revealed that phonons had a mean free path of ~28 μm in CVD-grown graphene, which makes phonons rather than electrons dominate the thermal conduction in graphene, leading to $K_e << K_p$ [59]. As GFs are composed of graphene, they are expected to inherit the high thermal conductivity of graphene. However, it is found that GFs exhibited thermal conductivity in a large range of 30–3300 W/m·K [25,26], which may be because of the following reasons:

1. Different fabrication methods: different fabrication methods produced GFs with different crystal structures. Top-down strategies of fabricating GFs from GO always induced lower thermal conductivity compared to those obtained from bottom-up strategies, which may produce different and clean crystal structures of the GFs.

2. Different reduction methods: in top-down strategies, different reduction methods, such as thermal annealing and chemical reduction, also result in different thermal conductivity of the GFs. Different temperatures in thermal reduction and different reducing agents in chemical reduction lead to different thermal conductivity.

3. Different post-treatment methods: GFs with or without compression or compressed under different pressure exhibited dissimilar thermal conductivity.

Yu and colleagues [26] modified GO with alkaline earth metal ions to fabricate GO films with tuned thermal conductivity. The GO films modified with Mg^{2+} and Ca^{2+} acquired enhanced thermal conductivity of 32.05 W/m·K and 61.38 W/m·K, respectively, which are more than 8 and 15 times that of crude GO films (3.91 W/m·K). Gee et al. prepared the GFs through an electrochemical exfoliation and filtration process [25]. The thermal conductivity of the as-prepared GFs was measured to be 3390 W/m·K by using the thermoelectric method. This is the highest reported thermal conductivity of GFs, which is even much higher than that of graphite (K_i = 2000 W/m·K) [58,60].

Peng et al. obtained the GO films through a scrape coating and direct evaporation [28]. After carbonized at 1300 °C and graphitized at 3000 °C, as well as compressed at 50–300 MPa, the rGO films exhibited a thermal conductivity as high as 1940 ± 113 W/m·K. This may be because (i) the high temperature (3000 °C) can reduce the functional groups on graphene (transforming the sp^3 crystal of GO to the sp^2 crystal of graphene), and (ii) the high temperature can also promote the self-healing of the defective graphene, inducing perfect graphene with sp^2 crystals. Both the functional group-free and defect-free crystals benefit from the thermal conduction in the GFs. Guo et al. combined blade-coating and direct evaporation to fabricate self-standing GO films [22]. After being reduced by vitamin C, the rGO films achieved a high thermal conductivity of 2600 W/m·K. For comparison, the thermal conductivity of the GFs synthesized using different methods is summarized in Table 1.

Table 1. Summary of the GFs obtained from different synthesis method.

Materials	Fabrication Method	Reduction Method/Post-Treatment	Thermal Measurement	Thermal Conductivity (W/m·K)
Graphene film [25]	Electrochemical exfoliation, vacuum filtration	–	Thermoelectric method	3300
rGO film [22]	Blade-coating, evaporation	Vitamin C reduction	Laser flash	2600
rGO film [28]	Scrape coating, evaporation	Annealed at 3000 °C, compressed at 50–300 MPa	Laser flash	1940
Graphene film [61]	Hydroxide-assisted exfoliation, vacuum filtration	Annealed at 2800 °C, compressed at 100 MPa	Laser flash	1842
rGO film [33]	Vacuum filtration	L-ascorbic acid reduction	Laser flash	1642
Graphene film [34]	Ball-milling, Filtration	Annealed at 2850 °C, compressed at 30 MPa	Self-heating	1529
rGO film [36]	Vacuum filtration	HI reduction	Laser flash	1390 ± 65
rGO film [62]	Vacuum filtration	Annealed at 1200 °C	Laser flash	1043.5
rGO film [42]	Evaporation	Annealed at 900 °C in 5% H_2-Ar gas	Laser flash	902
rGO film [63]	Roller coating	Annealed at 2800 °C	Laser flash	826
rGO film [48]	Filtration	Annealed at 1000 °C	Laser flash	373
Graphene nanoplatelet film [35]	Vacuum filtration	Annealed at 120 °C and 340 °C	Laser flash	313
rGO film [31]	Vacuum filtration	Annealed at 1060 °C	Angstrom method	220
rGO film [26]	Vacuum filtration	Metal ion modified	Laser flash	Mg-modified: 32.05 Ca-modified: 61.38
rGO film [38]	Direct evaporation	Annealed at 1000 °C	Laser flash	61

3.3. Parameters That Affect the Thermal Conductivity of Free-Standing Graphene Films

3.3.1. Thermal Annealing Temperature

GO has various functional groups and defects in its surface. Thermal annealing in a proper temperature may partially reduce GO to rGO, thus leading to the higher thermal conductivity of the GFs. Previous studies have corroborated that the thermal annealing temperature exerts a significant impact on the thermal conductivity of the GFs [28,37,38,40]. Renteria et al. [38] studied the effect of thermal annealing temperature on the thermal conductivity of the rGO films. The in-plane thermal conductivity of the rGO films dramatically increased from ~3 to ~61 W/m·K (room temperature) when GO films were annealed up to 1000 °C, as presented in Figure 7a. More recently, Peng et al. also found that the thermal conductivity of rGO films could be elevated by raising the annealing temperature [28]. The thermal conductivity of the rGO films increased from ~800 to ~2000 W/m·K (close to that of graphite) as the annealing temperature was raised from 1400 °C to 3000 °C, as displayed in Figure 7c.

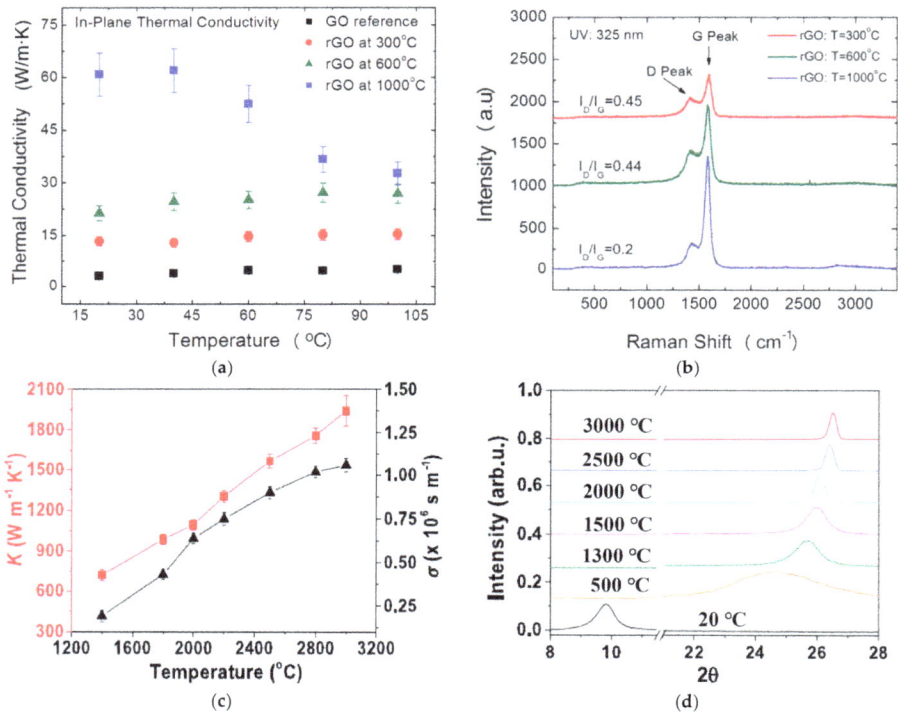

Figure 7. (**a**) Experimental in-plane thermal conductivity, *K*, as a function of temperature for rGO films annealed at different temperatures and a reference GO film. (**b**) Raman spectra of rGO films under UV (λ = 325 nm) laser excitations. The peaks at ≈1350 and 1580 cm^{-1} correspond to the D and G peaks, respectively. (**c**) The thermal (red line) and electrical conductivity (black line) of rGO films annealed at different temperatures. (**d**) XRD patterns of rGO films annealed at different temperatures. (**a,b**) Reprinted from [38] with permission, Copyright 2015 Wiley Online Library; (**c,d**) Reprinted from [28] with permission, Copyright 2017 Wiley Online Library.

The thermal conductivity increase with the rise of annealing temperature may be caused by the removal of the functional groups and defects of GO films, as well as the recovery of the sp^2 crystal structures from the sp^3 crystal. This was substantiated by the decreased I_D/I_G ratio in Raman spectroscopy and the different peaks in X-ray powder diffraction (XRD) patterns with raising the annealing temperature (Figure 7b,d). In addition to the thermal annealing temperature, the gas atmosphere during thermal annealing also affects the final thermal conductivity of the rGO films. Introducing a reducing gas, like H_2, could lead to a higher thermal conductivity of the rGO films. For instance, Liu et al. thermally annealed GO films at 900 °C in a 5% H_2-Ar gas mixture and the as-prepared rGO films obtained have a thermal conductivity as high as ~1200 W/m·K [64]—much higher than those annealed without H_2 at the same temperature [38].

3.3.2. The Lateral Size of Graphene and GO Sheets

Compared to electrons, phonons dominate the thermal conduction in graphene and related nanomaterials. Phonons exhibited a mean free path (MFP) as long as 28 μm in CVD-grown graphene, and the heat was found to be transferred through a ballistic mechanism in graphene and rGO [58,65]. The size of graphene or rGO has been demonstrated to greatly affect the thermal conductivity of the GFs in both experimental and theoretical studies [66–68], especially when the graphene size is close to or smaller than the MFP. Ma et al. tailored the thermal transport properties of graphene films by modulating the grain size of graphene [23]. They applied a segregation-adsorption CVD to grow graphene films with a tunable grain size from ~0.2 μm to ~10 μm. With increasing the grain size of graphene, the obtained GFs displayed a dramatic increase in thermal conductivity $(0.5~5.2 \times 10^3$ W/m·K). Peng et al. [28] also reported the obvious elevation from ~1000 W/m·K to ~1950 W/m·K for the thermal conductivity of the as-prepared debris-free graphene films (dfGFs) when the average size of the graphene sheets increased from ~5 μm to 100 μm.

Kumar et al. [36] fabricated small rGO films (rSGO) and a large-area rGO film (rLGO) and compared their thermal and electrical conductivity. The rSGO films exhibited a thermal conductivity of 900 W/m·K, while the rLGO films exhibited a thermal conductivity of 1390 W/m·K, indicating that higher thermal conductivity of rGO films can be achieved by utilizing large GO sheets. Reducing the size of graphene sheets not only shortens the mean free path of phonons in graphene sheets, but also increases the phonon scattering probability in the films, thus inducing lower thermal conductivity of the graphene films. The thermal conductivity of the graphene films may be turned by using the graphene or graphene oxide with a different size [36,69].

3.3.3. Hybridization with Other Components

To enhance the thermal conductivity of GFs, graphene or GO can be combined with other materials to form hybrids. Kong et al. randomly deposited carbon fiber (CF) as the porous scaffold on the porous metal plates through vacuum filtration [27]. Then GO hydrogel was subsequently deposited on the porous scaffold to form GO–CF composite (G–CF) films. After annealing at 1000 °C in Ar atmosphere, the as-prepared G-CF films exhibited a superior thermal conductivity of 977 W/m·K, significantly higher than those of graphene film (318 W/m·K), graphitized polymide (PI) film (743 W/m·K), and flexible graphite film (137 W/m·K). Hu et al. also combined a carbon nanotube (CNT) film with GO for hybrid films by applying CVD and a spray-coating method [54]. The as-prepared GO/CNT composite films were thermally annealed at 2800 °C to reduce GO to rGO. The rGO/CNT films delivered a high thermal conductivity of 1056 W/m·K, a superior mechanical strength of ~1 GPa, and an excellent electrical conductivity of 1182 S/cm.

Besides CNTs and CFs, other diverse materials have also been used in combination with graphene for advanced functional properties, such as polymers [70–72] and various 2D materials [73–75]. The graphene/polymer composite films generally attain a lower thermal conductivity compared to the pure graphene films. This may be due to the ultralow thermal conductivity of the polymer (~0.1 W/m·K) compared to that of graphene (>100 W/m·K), transforming the ballistic heat transfer in graphene to the diffusive heat transfer in polymers [76,77]. Moreover, a number of theoretical studies have demonstrated that the interfacial thermal resistance (often known as the Kapitza resistance) between graphene and polymer also lowers the thermal conductivity of the composite films [78–81].

The interfacial thermal resistance arises from poor thermal coupling between graphene and polymer, and can be significantly decreased by grafting the graphene sheets with polymer chains. However, excessive loading of grafted polymer may reduce the thermal conductivity of the graphene sheets, thus inducing lower effective thermal conductivity for the composite film. Therefore, the grafting density should be well-controlled to achieve a higher thermal conductivity of the composite film than the conductivity of the polymer matrix [81]. More discussion about nanoscale thermal transport based on phonon propagation can be found in other reviews [82,83]. When combining graphene and other nanomaterials with high thermal conductivity, like boron nitride (BN) or CNTs, the composite films still can achieve a desirable thermal conductivity. This may be due to low interfacial thermal resistances between graphene and other nanomaterials, owing to the strong phonon—phonon coupling [84,85]. Table 2 is a presentation of the typical hybrid films reported in the past three years, clearly manifesting the effects of different hybridization components on the thermal conductivity of the hybrid films.

Table 2. Summary of the graphene composite films combined with different materials.

Materials	Fabrication Method	Reduction Method/Post-Treatment	Thermal Measurement	Thermal Conductivity (W/m·K)
rGO/CNT film [54]	CVD, spray coating	Annealed at 2800 °C	Laser flash	1056
rGO/carbon fiber film [27]	Vacuum filtration	Annealed at 1000 °C	Laser flash	977
rGO/PBO film [86]	Dispersion, casting	120 °C to reduce GO	Laser flash	50
BN/GO film [87]	Vacuum filtration	–	Laser flash	29.8
GO/polymer/BN film [88]	casting	–	Laser flash	12.62
rGO/cellulose film [89]	Vacuum filtration	Hydrazine reduction	Laser flash	6.17
Graphene/PI film [90]	CVD/impregnation	–	Laser flash	3.73
Graphene/NRlatex film [91]	Ball milling dipping	–	Hot-disk	0.482
GO/MWNT films [92]	Vacuum filtration	–	Laser flash	0.35

3.3.4. Thickness and Density of the Graphene Films

The thickness of graphene films also plays a significant role in the thermal conductivity of graphene films. Zhang et al. compared the thermal conductivity of the GFs with different thicknesses from 20 to 60 µm [33]. They found that the in-plane thermal conductivity of the GFs decreased from 1642 W/m·K to 675 W/m·K as the thickness increased from 20 µm to 60 µm. This phenomenon was due to the enhancement of total defects in graphene films with bigger thickness. More defects would result in a more significant scattering of phonons, thus hindering the phonon propagation along the in-plane direction.

Not only the thickness, but also the density of the GFs would influence the thermal conductivity of the GFs. Xin et al. [40] conducted compaction to remove air pores to fabricate dense GFs, but with different density. Even at different annealed temperatures, the thermal conductivity of the GFs significantly elevated with the increase in the GFs' density. As shown in Figure 8, for the 2850 °C annealed GFs, when the density increased from 0.4 g/cm^3 to 2.0 g/cm^3, the thermal conductivity of the GFs increased from 200 W/m·K to ~1500 W/m·K. The significant increase of thermal conductivity of the GFs with the rise of density may be explained as described below:

i. Higher density means fewer air pores in the GFs, which may reduce the phonon scattering at air–graphene interfaces, thus inducing the higher thermal conductivity of the GFs.

ii. According to the equation ($K = \alpha\rho C_p$) used to calculate the thermal conductivity of the GFs, given thermal diffusivity α and specific capacity C_p, the thermal conductivity, K, increases linearly with density, ρ. As shown in Figure 8, the values of K show an approximately linear relationship with ρ.

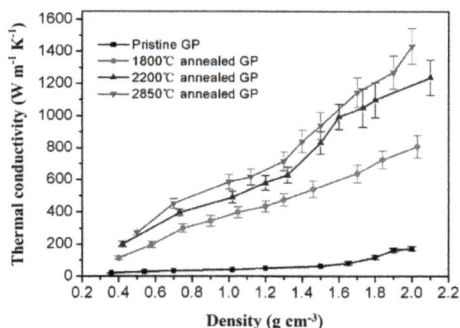

Figure 8. Change of thermal conductivity of free-standing graphene films with different density under different annealed temperatures. Reprinted with permission from [40]; Copyright 2014 Wiley Online Library.

4. Applications of Free-Standing Graphene Films in Thermal Engineering

The outstanding thermal conductivity of the GFs contributes to promising applications in diverse thermal management fields, such as thermal interface materials (TIMs), and heat dissipation materials (HDMs). For instance, to demonstrate the thermal management capabilities of their rGO films, Huang et al. applied a 7 W high-bright LED as a hot spot and recorded the temperature changes by a thermocouple sensor in the systems with and without rGO films [63], as illustrated in Figure 9a. The rGO films effectively reduced the temperature of the LED hot spot, 3 °C difference after few minutes shown in Figure 9b, confirming the heat dissipation ability of the as-prepared rGO films.

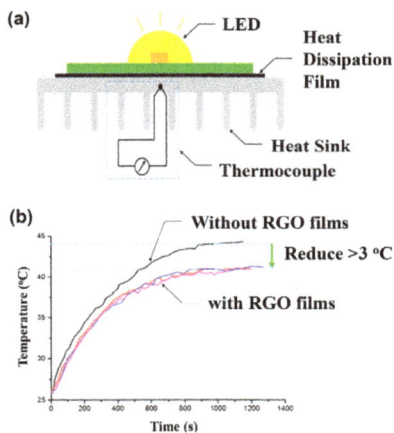

Figure 9. (a) Schematic plot of the temperature measurement model of a 7 W LED as the hot spot with and without rGO films. (b) Real-time temperature profile of the LED hot spot with and without rGO films. Reprinted from [63] with permission; Copyright 2017 Royal Society of Chemistry.

Guo et al. fabricated ultra-flexible, lightweight, and stretchable rGO films for wearable thermal management components [22]. They designed the rGO films into kirigami structures and integrated them into cloth for personal thermal management, as displayed in Figure 10. With a small applied voltage (3.2 V), a rapid heating-up response to 45 °C from room temperature could be achieved in 6 s. When the heating stopped, the rGO films cooled down to room temperature in 5 s, manifesting both fast electrical heating response and efficient heat dissipating capability. Zhang et al. also demonstrated the superb performance of the GFs as TIMs in microelectronic devices [33]. When combined with functional GO (FGO), the GFs exhibited a better TIMs performance due to the reduced thermal resistance between the GFs and the microelectronic chips by the FGO. Besides thermal management applications, GFs and related films can also be applied in various other applications, such as electromagnetic interference (EMI) shielding, gas barriers, and energy storage and sensors [35,36,88,93–100], owing to their superior electrochemical and mechanical properties.

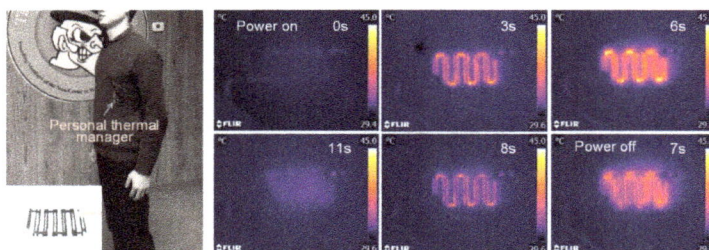

Figure 10. A demonstration of wearable rGO films for personal thermal management. The rGO films possessed kirigami structure and were integrated into cloth. A work cycle of heating–cooling can be implemented within 11 s. Reprinted from [22] with permission; Copyright 2017 Wiley Online Library.

5. Summary and Outlook

In this article, we have reviewed recent advances in the subject of thermal conductivity of free-standing GFs. Diverse "top-down" strategies, like vacuum filtration, direct evaporation, and various coating techniques, have been developed to prepare free-standing GFs with large areas. "Bottom-up" CVD methods can be used to synthesize GFs with high-quality crystals. For GFs obtained from the GO, thermal annealing is normally required to reduce the GO and even to graphitize the rGO films for high thermal conductivity. Higher annealing temperature generally results in heightened thermal conductivity of the GFs. The thermal conductivity of the GFs can be also enhanced by hybridizing with other conducting materials, decreasing the GF thickness, increasing the GFs density, as well as completely reducing the GO. The high thermal conductivity of the GFs enables them to be utilized in extensive thermal management applications, such as highly efficient heat-dissipating materials and thermal interfacial materials.

In existing studies, thermal annealing at high temperature, as high as 3000 °C, was carried out to reduce the GO and, thus, to heighten the thermal conductivity of the GFs. However, thermal annealing at such a high temperature is not practical, and is also an energy-consuming process, which may hinder the industrial manufacturing of GFs with this method. Therefore, new reduction approaches that are low-cost, environmentally friendly, and easily accessible should be the new research direction. In addition, a lot of previous investigations pursued the high thermal conductivity of the GFs rather than considering the realistic applications of GFs. For instance, few studies took into account the contact between the GFs and the hot spot substrates: ineffective contact between the GFs and substrates may induce high thermal contact resistance, thus significantly lowering the efficiency of the GFs. More effort should be made to integrate the GFs into the micro-/nano-electronic devices with compact size and low weight for broader applications of the GFs. Multi-scale theoretical modeling, like molecular dynamics simulations and finite element methods, can be involved to shed light on the heat transfer mechanisms in the GFs from the atomic scale to the macro-scale. More attention should also be paid to combining different techniques, either "top-down" or "bottom-up," to achieve the large-scale fabrication of the GFs for industrial applications, for example, acting as packaging materials in high-power battery packs for whole electrical vehicles.

Acknowledgments: We appreciate the financial support from the National Natural Science Foundation of China (51602038), the Sichuan Science and Technology Agency (2017HH0101, 2017GZ0113), and the Fundamental Research Funds for the Central Universities (ZYGX2016KYQD148).

Conflicts of Interest: The authors declare no conflicts of interest.

References

1. Zhu, H.; Li, Y.; Fang, Z.; Xu, J.; Cao, F.; Wan, J.; Preston, C.; Yang, B.; Hu, L. Highly thermally conductive papers with percolative layered boron nitride nanosheets. *ACS Nano* **2014**, *8*, 3606–3613. [CrossRef] [PubMed]

2. Chen, H.; Ginzburg, V.V.; Yang, J.; Yang, Y.; Liu, W.; Huang, Y.; Du, L.; Chen, B. Thermal conductivity of polymer-based composites: Fundamentals and applications. *Prog. Polym. Sci.* **2016**, *59*, 41–85. [CrossRef]

3. Moore, A.L.; Shi, L. Emerging challenges and materials for thermal management of electronics. *Mater. Today* **2014**, *17*, 163–174. [CrossRef]

4. Thermal Conductivity of common Materials and Gases. Available online: https://www.engineeringtoolbox.com/thermal-conductivity-d_429.html (accessed on 7 February 2018).

5. Hu, J.; Ruan, X.; Chen, Y.P. Thermal conductivity and thermal rectification in graphene nanoribbons: A molecular dynamics study. *Nano Lett.* **2009**, *9*, 2730–2735. [CrossRef] [PubMed]

6. Yan, Z.; Nika, D.L.; Balandin, A.A. Thermal properties of graphene and few-layer graphene: Applications in electronics. *IET Circuits Devices Syst.* **2015**, *9*, 4–12. [CrossRef]

7. Shahil, K.M.F.; Balandin, A.A. Graphene-multilayer graphene nanocomposites as highly efficient thermal interface materials. *Nano Lett.* **2012**, *12*, 861–867. [CrossRef] [PubMed]

8. Zhang, H.; Fonseca, A.F.; Cho, K. Tailoring thermal transport property of graphene through oxygen functionalization. *J. Phys. Chem. C* **2014**, *118*, 1436–1442. [CrossRef]

9. Shahil, K.M.F.; Balandin, A.A. Thermal properties of graphene and multilayer graphene: Applications in thermal interface materials. *Solid State Commun.* **2012**, *152*, 1331–1340. [CrossRef]

10. Gong, F.; Liu, X.; Yang, Y.; Xia, D.; Wang, W.; Duong, H.; Papavassiliou, D.; Xu, Z.; Liao, J.; Wu, M. A facile approach to tune the electrical and thermal properties of graphene aerogels by including bulk MoS$_2$. *Nanomaterials* **2017**, *7*, 420. [CrossRef] [PubMed]

11. Fan, Z.; Gong, F.; Nguyen, S.T.; Duong, H.M. Advanced multifunctional graphene aerogel–poly (methyl methacrylate) composites: Experiments and modeling. *Carbon* **2015**, *81*, 396–404. [CrossRef]

12. Fang, X.; Fan, L.-W.; Ding, Q.; Wang, X.; Yao, X.-L.; Hou, J.-F.; Yu, Z.-T.; Cheng, G.-H.; Hu, Y.-C.; Cen, K.-F. Increased thermal conductivity of eicosane-based composite phase change materials in the presence of graphene nanoplatelets. *Energy Fuels* **2013**, *27*, 4041–4047. [CrossRef]

13. Dai, W.; Yu, J.; Liu, Z.; Wang, Y.; Song, Y.; Lyu, J.; Bai, H.; Nishimura, K.; Jiang, N. Enhanced thermal conductivity and retained electrical insulation for polyimide composites with sic nanowires grown on graphene hybrid fillers. *Compos. Part A Appl. Sci. Manuf.* **2015**, *76*, 73–81. [CrossRef]

14. Qian, R.; Yu, J.; Wu, C.; Zhai, X.; Jiang, P. Alumina-coated graphene sheet hybrids for electrically insulating polymer composites with high thermal conductivity. *RSC Adv.* **2013**, *3*, 17373–17379. [CrossRef]

15. Yu, L.; Park, J.S.; Lim, Y.S.; Lee, C.S.; Shin, K.; Moon, H.J.; Yang, C.M.; Lee, Y.S.; Han, J.H. Carbon hybrid fillers composed of carbon nanotubes directly grown on graphene nanoplatelets for effective thermal conductivity in epoxy composites. *Nanotechnology* **2013**, *24*, 155604. [CrossRef] [PubMed]

16. Luan, V.H.; Tien, H.N.; Cuong, T.V.; Kong, B.-S.; Chung, J.S.; Kim, E.J.; Hur, S.H. Novel conductive epoxy composites composed of 2-D chemically reduced graphene and 1-D silver nanowire hybrid fillers. *J. Mater Chem.* **2012**, *22*, 8649–8653. [CrossRef]

17. Wang, F.; Drzal, L.T.; Qin, Y.; Huang, Z. Enhancement of fracture toughness, mechanical and thermal properties of rubber/epoxy composites by incorporation of graphene nanoplatelets. *Compos. Part A Appl. Sci. Manuf.* **2016**, *87*, 10–22. [CrossRef]

18. Han, Z.D.; Fina, A. Thermal conductivity of carbon nanotubes and their polymer nanocomposites: A review. *Prog. Polym. Sci.* **2011**, *36*, 914–944. [CrossRef]

19. Gong, F.; Duong, H.; Papavassiliou, D. Review of recent developments on using an off-lattice monte carlo approach to predict the effective thermal conductivity of composite systems with complex structures. *Nanomaterials* **2016**, *6*, 142. [CrossRef] [PubMed]

20. Bui, K.; Duong, H.M.; Striolo, A.; Papavassiliou, D.V. Effective heat transfer properties of graphene sheet nanocomposites and comparison to carbon nanotube nanocomposites. *J. Phys. Chem. C* **2011**, *115*, 3872–3880. [CrossRef]

21. Konatham, D.; Bui, K.N.D.; Papavassiliou, D.V.; Striolo, A. Simulation insights into thermally conductive graphene-based nanocomposites. *Mol. Phys.* **2011**, *109*, 97–111. [CrossRef]

22. Guo, Y.; Dun, C.; Xu, J.; Mu, J.; Li, P.; Gu, L.; Hou, C.; Hewitt, C.A.; Zhang, Q.; Li, Y.; et al. Ultrathin, washable, and large-area graphene papers for personal thermal management. *Small* **2017**, *13*, 1702645. [CrossRef] [PubMed]

23. Ma, T.; Liu, Z.; Wen, J.; Gao, Y.; Ren, X.; Chen, H.; Jin, C.; Ma, X.-L.; Xu, N.; Cheng, H.-M.; et al. Tailoring the thermal and electrical transport properties of graphene films by grain size engineering. *Nat. Commun.* **2017**, *8*, 14486. [CrossRef] [PubMed]

24. Xin, G.; Zhu, W.; Yao, T.; Scott, S.M.; Lian, J. Microstructure control of macroscopic graphene paper by electrospray deposition and its effect on thermal and electrical conductivities. *Appl. Phys. Lett.* **2017**, *110*, 091909. [CrossRef]

25. Gee, C.M.; Tseng, C.C.; Wu, F.Y.; Lin, C.T.; Chang, H.P.; Li, L.J.; Chen, J.C.; Hu, L.H. Few layer graphene paper from electrochemical process for heat conduction. *Mater. Res. Innov.* **2014**, *18*, 208–213. [CrossRef]

26. Yu, W.; Xie, H.; Li, F.; Zhao, J.; Zhang, Z. Significant thermal conductivity enhancement in graphene oxide papers modified with alkaline earth metal ions. *Appl. Phys. Lett.* **2013**, *103*, 141913. [CrossRef]

27. Kong, Q.-Q.; Liu, Z.; Gao, J.-G.; Chen, C.-M.; Zhang, Q.; Zhou, G.; Tao, Z.-C.; Zhang, X.-H.; Wang, M.-Z.; Li, F.; et al. Hierarchical graphene–carbon fiber composite paper as a flexible lateral heat spreader. *Adv Funct. Mater.* **2014**, *24*, 4222–4228. [CrossRef]

28. Peng, L.; Xu, Z.; Liu, Z.; Guo, Y.; Li, P.; Gao, C. Ultrahigh thermal conductive yet superflexible graphene films. *Adv. Mater.* **2017**, *29*, 1700589. [CrossRef] [PubMed]

29. Weng, Z.; Su, Y.; Wang, D.-W.; Li, F.; Du, J.; Cheng, H.-M. Graphene–cellulose paper flexible supercapacitors. *Adv. Energy Mater.* **2011**, *1*, 917–922. [CrossRef]

30. Wan, S.; Li, Y.; Peng, J.; Hu, H.; Cheng, Q.; Jiang, L. Synergistic toughening of graphene oxide–molybdenum disulfide–thermoplastic polyurethane ternary artificial nacre. *ACS Nano* **2015**, *9*, 708–714. [CrossRef] [PubMed]

31. Hou, Z.-L.; Song, W.-L.; Wang, P.; Meziani, M.J.; Kong, C.Y.; Anderson, A.; Maimaiti, H.; LeCroy, G.E.; Qian, H.; Sun, Y.-P. Flexible graphene–graphene composites of superior thermal and electrical transport properties. *ACS Appl. Mater. Interfaces* **2014**, *6*, 15026–15032. [CrossRef] [PubMed]

32. Huang, W.; Ouyang, X.; Lee, L.J. High-performance nanopapers based on benzenesulfonic functionalized graphenes. *ACS Nano* **2012**, *6*, 10178–10185. [CrossRef] [PubMed]

33. Zhang, Y.; Han, H.; Wang, N.; Zhang, P.; Fu, Y.; Murugesan, M.; Edwards, M.; Jeppson, K.; Volz, S.; Liu, J. Improved heat spreading performance of functionalized graphene in microelectronic device application. *Adv. Funct. Mater.* **2015**, *25*, 4430–4435. [CrossRef]

34. Teng, C.; Xie, D.; Wang, J.; Yang, Z.; Ren, G.; Zhu, Y. Ultrahigh conductive graphene paper based on ball-milling exfoliated graphene. *Adv. Funct. Mater.* **2017**, *27*, 1700240. [CrossRef]

35. Wu, H.; Drzal, L.T. Graphene nanoplatelet paper as a light-weight composite with excellent electrical and thermal conductivity and good gas barrier properties. *Carbon* **2012**, *50*, 1135–1145. [CrossRef]

36. Kumar, P.; Shahzad, F.; Yu, S.; Hong, S.M.; Kim, Y.-H.; Koo, C.M. Large-area reduced graphene oxide thin film with excellent thermal conductivity and electromagnetic interference shielding effectiveness. *Carbon* **2015**, *94*, 494–500. [CrossRef]

37. Shen, B.; Zhai, W.; Zheng, W. Ultrathin flexible graphene film: An excellent thermal conducting material with efficient EMI shielding. *Adv. Funct. Mater.* **2014**, *24*, 4542–4548. [CrossRef]

38. Renteria, J.D.; Ramirez, S.; Malekpour, H.; Alonso, B.; Centeno, A.; Zurutuza, A.; Cocemasov, A.I.; Nika, D.L.; Balandin, A.A. Strongly anisotropic thermal conductivity of free-standing reduced graphene oxide films annealed at high temperature. *Adv. Funct. Mater.* **2015**, *25*, 4664–4672. [CrossRef]

39. Chen, C.; Yang, Q.-H.; Yang, Y.; Lv, W.; Wen, Y.; Hou, P.-X.; Wang, M.; Cheng, H.-M. Self-assembled free-standing graphite oxide membrane. *Adv. Mater.* **2009**, *21*, 3007–3011. [CrossRef]

40. Xin, G.; Sun, H.; Hu, T.; Fard, H.R.; Sun, X.; Koratkar, N.; Borca-Tasciuc, T.; Lian, J. Large-area freestanding graphene paper for superior thermal management. *Adv. Mater.* **2014**, *26*, 4521–4526. [CrossRef] [PubMed]

41. Zhu, Y.; Murali, S.; Cai, W.; Li, X.; Suk, J.W.; Potts, J.R.; Ruoff, R.S. Graphene and graphene oxide: Synthesis, properties, and applications. *Adv. Mater.* **2010**, *22*, 3906–3924. [CrossRef] [PubMed]

42. Huang, S.-Y.; Zhao, B.; Zhang, K.; Yuen, M.M.F.; Xu, J.-B.; Fu, X.-Z.; Sun, R.; Wong, C.-P. Enhanced reduction of graphene oxide on recyclable cu foils to fabricate graphene films with superior thermal conductivity. *Sci. Rep.* **2015**, *5*, 14260. [CrossRef] [PubMed]

43. Fan, Z.; Marconnet, A.; Nguyen, S.T.; Lim, C.Y.H.; Duong, H.M. Effects of heat treatment on the thermal properties of highly nanoporous graphene aerogels using the infrared microscopy technique. *Int. J. Heat Mass Transf.* **2014**, *76*, 122–127. [CrossRef]

44. Fan, Z.; Tng, D.Z.Y.; Lim, C.X.T.; Liu, P.; Nguyen, S.T.; Xiao, P.; Marconnet, A.; Lim, C.Y.H.; Duong, H.M. Thermal and electrical properties of graphene/carbon nanotube aerogels. *Colloids Surf. A Physicochem. Eng. Asp.* **2014**, *445*, 48–53. [CrossRef]

45. Fan, Z.; Tng, D.Z.Y.; Nguyen, S.T.; Feng, J.D.; Lin, C.F.; Xiao, P.F.; Lu, L.; Duong, H.M. Morphology effects on electrical and thermal properties of binderless graphene aerogels. *Chem. Phys. Lett.* **2013**, *561*, 92–96. [CrossRef]

46. Li, A.; Zhang, C.; Zhang, Y.-F. Thermal conductivities of PU composites with graphene aerogels reduced by different methods. *Compos. Part A Appl. Sci. Manuf.* **2017**, *103*, 161–167. [CrossRef]

47. Yang, W.; Zhao, Z.; Wu, K.; Huang, R.; Liu, T.; Jiang, H.; Chen, F.; Fu, Q. Ultrathin flexible reduced graphene oxide/cellulose nanofiber composite films with strongly anisotropic thermal conductivity and efficient electromagnetic interference shielding. *J. Mater. Chem. C* **2017**, *5*, 3748–3756. [CrossRef]

48. Jin, S.; Gao, Q.; Zeng, X.; Zhang, R.; Liu, K.; Shao, X.; Jin, M. Effects of reduction methods on the structure and thermal conductivity of free-standing reduced graphene oxide films. *Diam. Relat. Mater.* **2015**, *58*, 54–61. [CrossRef]

49. Pham, V.H.; Cuong, T.V.; Hur, S.H.; Shin, E.W.; Kim, J.S.; Chung, J.S.; Kim, E.J. Fast and simple fabrication of a large transparent chemically-converted graphene film by spray-coating. *Carbon* **2010**, *48*, 1945–1951. [CrossRef]

50. Gilje, S.; Han, S.; Wang, M.; Wang, K.L.; Kaner, R.B. A chemical route to graphene for device applications. *Nano Lett.* **2007**, *7*, 3394–3398. [CrossRef] [PubMed]

51. Zhao, J.; Pei, S.; Ren, W.; Gao, L.; Cheng, H.-M. Efficient preparation of large-area graphene oxide sheets for transparent conductive films. *ACS Nano* **2010**, *4*, 5245–5252. [CrossRef] [PubMed]

52. Wu, J.; Becerril, H.A.; Bao, Z.; Liu, Z.; Chen, Y.; Peumans, P. Organic solar cells with solution-processed graphene transparent electrodes. *Appl. Phys. Lett.* **2008**, *92*, 263302. [CrossRef]

53. Polsen, E.S.; McNerny, D.Q.; Viswanath, B.; Pattinson, S.W.; Hart, A.J. High-speed roll-to-roll manufacturing of graphene using a concentric tube CVD reactor. *Sci. Rep.* **2015**, *5*, 10257. [CrossRef] [PubMed]

54. Hu, D.; Gong, W.; Di, J.; Li, D.; Li, R.; Lu, W.; Gu, B.; Sun, B.; Li, Q. Strong graphene-interlayered carbon nanotube films with high thermal conductivity. *Carbon* **2017**, *118*, 659–665. [CrossRef]

55. Novoselov, K.S.; Falko, V.I.; Colombo, L.; Gellert, P.R.; Schwab, M.G.; Kim, K. A roadmap for graphene. *Nature* **2012**, *490*, 192–200. [CrossRef] [PubMed]

56. Yang, J.; Qi, G.-Q.; Liu, Y.; Bao, R.-Y.; Liu, Z.-Y.; Yang, W.; Xie, B.-H.; Yang, M.-B. Hybrid graphene aerogels/phase change material composites: Thermal conductivity, shape-stabilization and light-to-thermal energy storage. *Carbon* **2016**, *100*, 693–702. [CrossRef]

57. Balandin, A.A. Thermal properties of graphene and nanostructured carbon materials. *Nat. Mater.* **2011**, *10*, 569–581. [CrossRef] [PubMed]

58. Balandin, A.A.; Ghosh, S.; Bao, W.; Calizo, I.; Teweldebrhan, D.; Miao, F.; Lau, C.N. Superior thermal conductivity of single-layer graphene. *Nano Lett.* **2008**, *8*, 902–907. [CrossRef] [PubMed]

59. Banszerus, L.; Schmitz, M.; Engels, S.; Goldsche, M.; Watanabe, K.; Taniguchi, T.; Beschoten, B.; Stampfer, C. Ballistic transport exceeding 28 μm in CVD grown graphene. *Nano Lett.* **2015**, *16*, 1387. [CrossRef] [PubMed]

60. Fugallo, G.; Cepellotti, A.; Paulatto, L.; Lazzeri, M.; Marzari, N.; Mauri, F. Thermal conductivity of graphene and graphite: Collective excitations and mean free paths. *Nano Lett.* **2014**, *14*, 6109–6114. [CrossRef] [PubMed]

61. Ding, J.; ur Rahman, O.; Zhao, H.; Peng, W.; Dou, H.; Chen, H.; Yu, H. Hydroxylated graphene-based flexible carbon film with ultrahigh electrical and thermal conductivity. *Nanotechnology* **2017**, *28*, 39LT01. [CrossRef] [PubMed]

62. Song, N.-J.; Chen, C.-M.; Lu, C.; Liu, Z.; Kong, Q.-Q.; Cai, R. Thermally reduced graphene oxide films as flexible lateral heat spreaders. *J. Mater. Chem. A* **2014**, *2*, 16563–16568. [CrossRef]

63. Huang, Y.; Gong, Q.; Zhang, Q.; Shao, Y.; Wang, J.; Jiang, Y.; Zhao, M.; Zhuang, D.; Liang, J. Fabrication and molecular dynamics analyses of highly thermal conductive reduced graphene oxide films at ultra-high temperatures. *Nanoscale* **2017**, *9*, 2340–2347. [CrossRef] [PubMed]

64. Liu, S.Q.; Zhang, K.; Yuen, M.M.F.; Fu, X.Z.; Sun, R.; Wong, C.P. Effect of reduction temperatures on the thermal and electrical conductivities of reduced graphene oxide films on the CU foils. In Proceedings of the 2016 17th International Conference on Electronic Packaging Technology (ICEPT), Guangzhou, China, 16–19 August 2016; pp. 310–312.

65. Pop, E.; Varshney, V.; Roy, A.K. Thermal properties of graphene: Fundamentals and applications. *MRS Bull.* **2012**, *37*, 1273–1281. [CrossRef]

66. Ng, T.Y.; Yeo, J.J.; Liu, Z.S. A molecular dynamics study of the thermal conductivity of graphene nanoribbons containing dispersed stone–thrower–wales defects. *Carbon* **2012**, *50*, 4887–4893. [CrossRef]

67. Cui, L.; Du, X.; Wei, G.; Feng, Y. Thermal conductivity of graphene wrinkles: A molecular dynamics simulation. *J. Phys. Chem. C* **2016**, *120*, 23807–23812. [CrossRef]

68. Xu, W.; Zhang, G.; Li, B. Thermal conductivity of penta-graphene from molecular dynamics study. *J. Chem. Phys.* **2015**, *143*, 154703. [CrossRef] [PubMed]

69. Shtein, M.; Nadiv, R.; Buzaglo, M.; Kahil, K.; Regev, O. Thermally conductive graphene-polymer composites: Size, percolation, and synergy effects. *Chem. Mater.* **2015**, *27*, 2100–2106. [CrossRef]

70. Song, W.-L.; Cao, M.-S.; Lu, M.-M.; Bi, S.; Wang, C.-Y.; Liu, J.; Yuan, J.; Fan, L.-Z. Flexible graphene/polymer composite films in sandwich structures for effective electromagnetic interference shielding. *Carbon* **2014**, *66*, 67–76. [CrossRef]

71. Kumar, P.; Kumar, A.; Cho, K.Y.; Das, T.K.; Sudarsan, V. An asymmetric electrically conducting self-aligned graphene/polymer composite thin film for efficient electromagnetic interference shielding. *AIP Adv.* **2017**, *7*, 015103. [CrossRef]

72. Rahim, J.; Amir, H.; Muhammad Aftab, A.; Imtiaz, A.; Attaullah, S.; Muhammad, S.; Akhtar, H. Flexible, thin films of graphene–polymer composites for emi shielding. *Mater. Res. Express* **2017**, *4*, 035605.

73. Wang, B.; Zhang, Y.; Zhang, J.; Xia, R.; Chu, Y.; Zhou, J.; Yang, X.; Huang, J. Facile synthesis of a MoS_2 and functionalized graphene heterostructure for enhanced lithium-storage performance. *ACS Appl. Mater. Interfaces* **2017**, *9*, 12907–12913. [CrossRef] [PubMed]

74. Fang, Y.; Lv, Y.; Gong, F.; Elzatahry, A.A.; Zheng, G.; Zhao, D. Synthesis of 2D-mesoporous-carbon/ MoS_2 heterostructures with well-defined interfaces for high-performance lithium-ion batteries. *Adv. Mater.* **2016**, *28*, 9385–9390. [CrossRef] [PubMed]

75. Teng, Y.; Zhao, H.; Zhang, Z.; Li, Z.; Xia, Q.; Zhang, Y.; Zhao, L.; Du, X.; Du, Z.; Lv, P.; et al. MoS_2 nanosheets vertically grown on graphene sheets for lithium-ion battery anodes. *ACS Nano* **2016**, *10*, 8526–8535. [CrossRef] [PubMed]

76. Gong, F.; Papavassiliou, D.V.; Duong, H.M. Off-lattice monte carlo simulation of heat transfer through carbon nanotube multiphase systems taking into account thermal boundary resistances. *Numer. Heat Transf. Part A Appl.* **2014**, *65*, 1023–1043. [CrossRef]

77. Gong, F.; Duong, H.M.; Papavassiliou, D.V. Inter-carbon nanotube contact and thermal resistances in heat transport of three-phase composites. *J. Phys. Chem. C* **2015**, *119*, 7614–7620. [CrossRef]

78. Eslami, H.; Mohammadzadeh, L.; Mehdipour, N. Reverse nonequilibrium molecular dynamics simulation of thermal conductivity in nanoconfined polyamide-6,6. *J. Chem. Phys.* **2011**, *135*, 064703. [CrossRef] [PubMed]

79. Eslami, H.; Mohammadzadeh, L.; Mehdipour, N. Anisotropic heat transport in nanoconfined polyamide-6,6 oligomers: Atomistic reverse nonequilibrium molecular dynamics simulation. *J. Chem. Phys.* **2012**, *136*, 104901. [CrossRef] [PubMed]

80. Eslami, H.; Mehdipour, F.; Setoodeh, A.; Rouzegar, J. Nanoconfined polymers: Modelling and simulation approaches. *Mol. Simul.* **2015**, *41*, 367–381. [CrossRef]

81. Gao, Y.; Müller-Plathe, F. Increasing the thermal conductivity of graphene-polyamide-6,6 nanocomposites by surface-grafted polymer chains: Calculation with molecular dynamics and effective-medium approximation. *J. Phys. Chem. B* **2016**, *120*, 1336–1346. [CrossRef] [PubMed]

82. Marconnet, A.M.; Ashegi, M.; Goodson, K.E. From the casimir limit to phononic crystals: 20 years of phonon transport studies using silicon-on-insulator technology. *J. Heat Transf.* **2013**, *135*, 061601. [CrossRef]

83. Marconnet, A.M.; Panzer, M.A.; Goodson, K.E. Thermal conduction phenomena in carbon nanotubes and related nanostructured materials. *Rev. Mod. Phys.* **2013**, *85*, 1295–1326. [CrossRef]

84. Gong, F.; Liu, J.; Yang, J.; Qin, J.; Yang, Y.; Feng, T.; Liu, W.; Duong, H.; Papavassiliou, D.V.; Wu, M. Effective thermal transport properties in multiphase biological systems containing carbon nanomaterials. *RSC Adv.* **2017**, *7*, 13615–13622. [CrossRef]

85. Lin, S.C.; Buehler, M.J. The effect of non-covalent functionalization on the thermal conductance of graphene/organic interfaces. *Nanotechnology* **2013**, *24*, 165702. [CrossRef] [PubMed]

86. Zhao, Y.; Kong, J.; Liu, H.; Zhuang, Q.; Gu, J.; Guo, Z. Ultra-high thermally conductive and rapid heat responsive poly(benzobisoxazole) nanocomposites with self-aligned graphene. *Nanoscale* **2016**, *8*, 19984–19993. [CrossRef] [PubMed]

87. Yao, Y.; Zeng, X.; Wang, F.; Sun, R.; Xu, J.-B.; Wong, C.-P. Significant enhancement of thermal conductivity in bioinspired freestanding boron nitride papers filled with graphene oxide. *Chem. Mater.* **2016**, *28*, 1049–1057. [CrossRef]

88. Zhang, X.; Zhang, X.; Yang, M.; Yang, S.; Wu, H.; Guo, S.; Wang, Y. Ordered multilayer film of (graphene oxide/polymer and boron nitride/polymer) nanocomposites: An ideal emi shielding material with excellent electrical insulation and high thermal conductivity. *Compos. Sci. Technol.* **2016**, *136*, 104–110. [CrossRef]

89. Song, N.; Jiao, D.; Ding, P.; Cui, S.; Tang, S.; Shi, L. Anisotropic thermally conductive flexible films based on nanofibrillated cellulose and aligned graphene nanosheets. *J. Mater. Chem. C* **2016**, *4*, 305–314. [CrossRef]

90. Gong, J.; Liu, Z.; Yu, J.; Dai, D.; Dai, W.; Du, S.; Li, C.; Jiang, N.; Zhan, Z.; Lin, C.-T. Graphene woven fabric-reinforced polyimide films with enhanced and anisotropic thermal conductivity. *Compos. Part A Appl. Sci. Manuf.* **2016**, *87*, 290–296. [CrossRef]

91. George, G.; Sisupal, S.B.; Tomy, T.; Pottammal, B.A.; Kumaran, A.; Suvekbala, V.; Gopimohan, R.; Sivaram, S.; Ragupathy, L. Thermally conductive thin films derived from defect free graphene-natural rubber latex nanocomposite: Preparation and properties. *Carbon* **2017**, *119*, 527–534. [CrossRef] [PubMed]

92. Hwang, Y.; Kim, M.; Kim, J. Enhancement of thermal and mechanical properties of flexible graphene oxide/carbon nanotube hybrid films though direct covalent bonding. *J. Mater. Sci.* **2013**, *48*, 7011–7021. [CrossRef]

93. Yu, A.; Roes, I.; Davies, A.; Chen, Z. Ultrathin, transparent, and flexible graphene films for supercapacitor application. *Appl. Phys. Lett.* **2010**, *96*, 253105. [CrossRef]

94. Obreja, A.C.; Iordanescu, S.; Gavrila, R.; Dinescu, A.; Comanescu, F.; Matei, A.; Danila, M.; Dragoman, M.; Iovu, H. Flexible films based on graphene/polymer nanocomposite with improved electromagnetic interference shielding. In Proceedings of the 2015 International Semiconductor Conference (CAS), Sinaia, Romania, 12–14 October 2015; pp. 49–52.

95. Indrani, B.; Tsegie, F.; Zlatka, S.; Paul, G.H.; Chen, J.; Ashwani, K.S.; Asim, K.R. Graphene films printable on flexible substrates for sensor applications. *2D Mater.* **2017**, *4*, 015036.

96. Pierleoni, D.; Xia, Z.Y.; Christian, M.; Ligi, S.; Minelli, M.; Morandi, V.; Doghieri, F.; Palermo, V. Graphene-based coatings on polymer films for gas barrier applications. *Carbon* **2016**, *96*, 503–512. [CrossRef]

97. Veronese, G.P.; Allegrezza, M.; Canino, M.; Centurioni, E.; Ortolani, L.; Rizzoli, R.; Morandi, V.; Summonte, C. Graphene as transparent conducting layer for high temperature thin film device applications. *Sol. Energy Mater. Sol. Cells* **2015**, *138*, 35–40. [CrossRef]

98. Song, W.; Kim, K.W.; Chang, S.-J.; Park, T.J.; Kim, S.H.; Jung, M.W.; Lee, G.; Myung, S.; Lim, J.; Lee, S.S.; et al. Direct growth of graphene nanopatches on graphene sheets for highly conductive thin film applications. *J. Mater. Chem. C* **2015**, *3*, 725–728. [CrossRef]

99. Anandan, S.; Narasinga Rao, T.; Sathish, M.; Rangappa, D.; Honma, I.; Miyauchi, M. Superhydrophilic graphene-loaded tio2 thin film for self-cleaning applications. *ACS Appl. Mater. Interfaces* **2013**, *5*, 207–212. [CrossRef] [PubMed]

100. Mosciatti, T.; Haar, S.; Liscio, F.; Ciesielski, A.; Orgiu, E.; Samorì, P. A multifunctional polymer-graphene thin-film transistor with tunable transport regimes. *ACS Nano* **2015**, *9*, 2357–2367. [CrossRef] [PubMed]

coatings

MDPI

Review

2D Materials-Coated Plasmonic Structures for SERS Applications

Ming Xia

Applied Materials Inc., Santa Clara, CA 95054, USA; xiaming@g.ucla.edu

Received: 27 February 2018; Accepted: 10 April 2018; Published: 12 April 2018

Abstract: Two-dimensional (2D) materials, such as graphene and hexagonal boron nitride, are new kinds of materials that can serve as substrates for surface enhanced Raman spectroscopy (SERS). When combined with traditional metallic plasmonic structures, the hybrid 2D materials/metal SERS platform brings extra benefits, including higher SERS enhancement factors, oxidation protection of the metal surface, and protection of molecules from photo-induced damages. This review paper gives an overview of recent progress in the 2D materials-coated plasmonic structure in SERS application, focusing on the fabrication of the hybrid 2D materials/metal SERS platform and its applications for Raman enhancement.

Keywords: surface enhanced Raman spectroscopy; two-dimensional materials; plasmonic structure

1. Introduction

Raman spectroscopy is an optical analysis technique providing characteristic spectral information of anlaytes and has a wide variety of applications in chemistry, biology, and medicine [1–4] because of its capability of providing fingerprints of molecule vibration. One major drawback of Raman spectroscopy is the low yield of Raman scattering, leading to weak Raman signals in most cases, and thus ordinary Raman spectroscopy can hardly provide discernable signals of a trace amount of analytes. Surface-enhanced Raman spectroscopy (SERS) makes up this deficiency via plasmon resonance from metallic nanostructures. Molecules adsorbed on the nanostructured metallic surface experience a large amplification of the electromagnetic (EM) field due to local surface plasmon resonance, which leads to an orders of magnitude increase in Raman yield and greatly enhanced Raman signal. SERS is capable of ultra-sensitive detection (single molecule detection) and allows for label-free detection with a high degree of specificity [5–9]. To achieve high SERS enhancement factors, many efforts have been devoted to developing various metallic (mainly Au and Ag) nanostructures to enhance the local EM field [10–16]. In addition to the bare metal SERS structure, a hybrid structure comprised of metal/inorganic materials [14,17–20] is also employed as a SERS sensor.

Two-dimensional (2D) materials, such as graphene and hexagonal boron nitride (h-BN), have unique electronic and optical properties, and attract widespread interest in potential applications in electronic devices, sensors, and energy generation [21–25]. In addition, 2D materials have been explored to enhance Raman signals [26–32]. Since the discovery of graphene's Raman enhancement capability [33], extensive research has been conducted to reveal the enhancing mechanism of two-dimensional materials, as well as their application in Raman enhancement substrates [11,13,30,34–36]. Unlike traditional SERS substrates, 2D materials provide a non-metallic surface with which to enhance the Raman signal. Recently, combining 2D materials with metallic plasmonic structures to form a hybrid SERS platform has become an emerging research field. The 2D materials-coated SERS platform offers synergetic Raman enhancement from both 2D materials and plasmon resonance, and additional advantages such as protection of metal from oxidation and protection of molecules from photo-induced damages. This paper will first give a brief introduction of the Raman enhancing mechanism of 2D materials and then discuss the recent process of 2D materials-coated plasmonic structures for SERS application, including their fabrication, sensitivity, and stability.

2. Raman Enhancement of 2D Materials

This section will briefly introduce the Raman enhancement mechanism of 2D materials, including graphene, h-BN, and molybdenum disulfide (MoS_2). Unlike EM enhancement mechanism of most metallic SERS substrates, Raman enhancement of 2D materials is due to chemical enhancement mechanism [26,33,37,38]. Chemical enhancement factor on metallic surface is usually low (~10–100) [39] compared with EM enhancement factor (~10^6–10^{11}) [5,40,41]. From a broad perspective, chemical enhancement can be considered as modification of the Raman polarizability tensor of molecule upon its adsorption, which in turn enhances or quenches Raman signals of vibrational modes [42,43]. In normal Raman scattering process, molecules are excited by external light to a high-energy level (an intermediate virtual state), and then molecules relax to ground state and emit Raman scattered photons. If the energy of the intermediate virtual state happens to be the same as one of the real electronic levels in the molecule, this scattering process is called resonance Raman scattering, which will have higher scattering efficiency and enhanced Raman signal. Charge-transfer model is often employed to explain chemical enhancement mechanism when molecules are adsorbed on metallic surface. One scenario of charge-transfer mechanisms is that molecules and metal form a surface complex by chemical bonding, which may cause a substantial change in the intrinsic polarizability of the molecule. This new surface complex creates a new electronic state, which is in resonance with the laser and shows enhanced Raman signal. This charge transfer model is the so-called excited state charge transfer model [44]. 2D materials provide a superior platform to study the chemical enhancement mechanism, because they have no dangling bonds in vertical direction and have atomically flat surface, and thus offer a pure system for the study of chemical enhancement effect.

Graphene is the first 2D material used to enhance Raman signals of molecules [33]. Raman enhancement of pristine graphene is ascribed to the ground state charge transfer mechanism [28,37] instead of the aforementioned excited state charge transfer mechanism. In ground state charge transfer process, analyte molecules do not form chemical bond with SERS substrate necessarily, and charge transfer happens when the substrate and the molecules are in the ground state. The charge transfer between molecules and graphene is a physical interaction instead of chemical bonding formation, and thus causes minor change in analytes' electronic distribution. Ground-state charge transfer can easily happen between graphene and molecules adsorbed on its surface because of graphene's two unique features: abundant π electrons on its surface and continuous energy band. Figure 1a shows the proposed ground state charge-transfer process between a dye molecule and graphene. In this process, the graphene electrons involvement in the Raman scattered process can enhance the electron−phonon coupling and thus induce the enhancement of the Raman signals [37]. It has been found that graphene Raman enhancement is vibration-mode dependent. The vibrational mode involving the lone pair or π electrons, which has stronger coupling with graphene, has highest Raman enhancement [29,45]. A more in-depth explanation of graphene-based surface enhancement scattering (GERS) has been given in [37].

h-BN and MoS_2 are other two kinds of 2D materials with different electronic and optical properties from graphene. h-BN is highly polarized and insulating with a large band gap of 5.9 eV [46]. CuPc molecule Raman signal is found to be enhanced by h-BN substrate. One proposed Raman enhancement mechanism of h-BN is the interface dipole interaction with analyte molecules, which causes symmetry-related perturbation in the CuPc molecule [26]. In addition, the Raman enhancement factor does not depend on the h-BN layer thickness, because the distribution of the intensity is uniform no matter how thick the h-BN flake is. Atomic layer thin MoS_2 is semiconductor and also has a polar bond [47]. For MoS_2, both the charge transfer and interface dipole interaction contribute to the Raman enhancement, but both contributions are much weaker compared with graphene and h-BN, respectively. The Raman enhancement of MoS$_2$ is not as obvious as that of graphene and h-BN, as shown in Figure 1b.

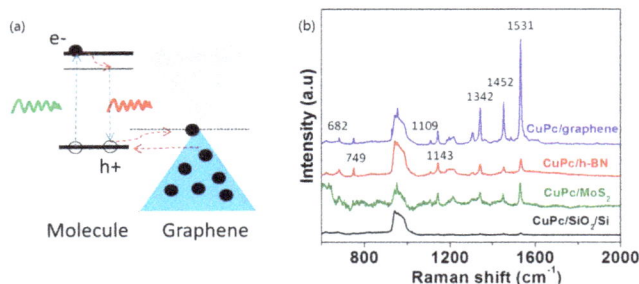

Figure 1. (**a**) Schematic of the Raman scattered process of graphene-enhanced Raman spectroscopy. Reproduced from [37] with permission; Copyright ACS 2012. (**b**) Raman spectra of the CuPc molecule on the blank SiO$_2$/Si substrate, on graphene, on h-BN, and on MoS$_2$ substrates. The numbers marked on the peaks are the peak frequencies of the Raman signals from the CuPc molecule. Reproduced from [26] with permission; Copyright ACS 2014.

3. Two Dimensional Materials-Coated Plasmonic Nanostructures

Traditional SERS analysis relies on metallic nanostructures that can generate strong local EM field. When combining 2D materials with metallic structure, the hybrid SERS substrate can provide even higher SERS enhancement factor due to the synergic effect of electromagnetic and chemical enhancement. 2D materials, like graphene and h-BN, could offer chemically inert and biocompatible surfaces [48–50], which is favorable in bio-detection. With 2D materials as shielding layer on metallic surface, metallic SERS platforms such as Ag could be protected from oxidation and have longer shelf life, which can improve the stability and repeatability of SERS analysis. The following discussion will focus on the fabrication, sensitivity, and stability of 2D materials/plasmonic structure for SERS application.

3.1. Fabrication

2D materials/plasmonic structures require incorporation of 2D materials and metallic plasmonic structures that can provide high local electric field upon laser excitation. Common fabrication methods of 2D materials include mechanical exfoliation, chemical exfoliation, and chemical vapor deposition (CVD). Summary of 2D materials synthesis [51–53] and metallic SERS substrate [54–56] fabrication can be found elsewhere. This section will focus on the incorporation of 2D materials with metallic plasmonic structure.

One simple way to incorporate 2D materials with metallic nanostructure is to transfer CVD grown 2D materials on metal surface. Graphene and MoS$_2$ have been proven to be capable of overlapping on Au nanostructures and generating strong Raman signals of graphene and MoS$_2$ [57]. Zhu et al. [58] fabricated graphene-covered gold nanovoid arrays using CVD grown monolayer graphene and investigated the SERS performance of graphene/plasmonic structure. Figure 2 shows the graphene transfer process and the SEM images of graphene-covered gold nanovoid arrays. In this study, graphene was actually suspended on Au nanovoid arrays instead of being conformally coated on Au surface. To achieve 2D material conformally coated SERS substrates, metallic structures need to have certain morphology. For instance, nanopyramid and nanocone structure can be conformally coated with 2D materials, although some ripples are unavoidable. Figure 3 shows graphene-coated Au nanopyramid [30] and MoS$_2$-coated SiO$_2$ nanocone [59], in which 2D materials are transferred with the assistance of poly(methyl methacrylate) (PMMA). Metallic plasmonic structure with conformally coated 2D materials can be better isolated from air and thus has longer stability.

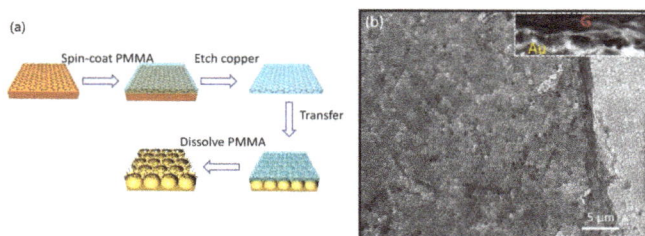

Figure 2. (**a**) Schematic illustrations of the graphene transfer process. (**b**) SEM image of a large-area nanovoid array integrated with the transferred monolayer graphene. The dark region is covered by graphene. The inset shows a SEM image of the cross-section of graphene-covered nanovoids. Reproduced from [58] with permission; Copyright ACS 2013.

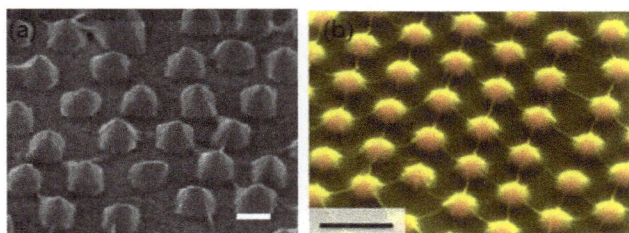

Figure 3. (**a**) Graphene-coated Au nanopyramid structure. Scale bare is 200 nm. Reproduced from [30] with permission; Copyright ACS 2015. (**b**) Tilted false-colour SEM image of the 2D strained MoS_2 crystal defined by the nanocone array. Scale bar is 500 nm. Reproduced from [59] with permission; Copyright Springer 2015.

PMMA-assisted transfer method has advantage of being able to coat 2D materials for plasmonic structures with various morphologies. However, the drawback is that PMMA residue left on the surface of 2D materials [60] may generate noisy Raman peaks and prevent analyte molecules directly adsorbed on the surface of 2D materials. Therefore, special care needs to be paid to avoid large amount of PMMA residue. Another concern of this transfer method is that the capillary force during the drying process of 2D materials may tear apart the 2D materials and expose the metallic surface to ambient environment. Xu et al. [27] developed a novel flexible graphene/plasmonic structure with PMMA as a carrying substrate for SERS application. In this study, PMMA was used to support a flat graphene surface instead of a sacrificing transfer layer. Figure 4 shows the fabrication process of the flexible graphene SERS tape.

Another way to incorporate 2D materials with metallic nanostructure is to use chemically exfoliated 2D materials to coat metallic nanoparticles. Kim et al. [61] developed a method with which to sandwich Ag nanoparticles between layers of reduced graphene oxide (rGO) and graphene oxide (GO) in order to prevent Ag nanoparticle from oxidation and boost Raman signals of analytes. Compared with CVD grown 2D materials, chemically exfoliated 2D materials are cost effective and easily to functionalize [61–64]. Figure 5 shows the preparation of SERS substrates with chemically exfoliated graphene.

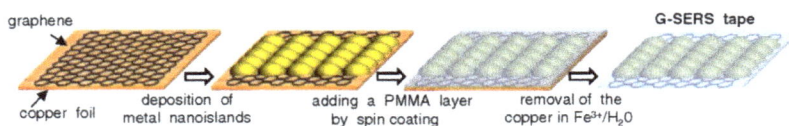

Figure 4. Schematic steps of the preparation route flexible G-SERS tape prepared from CVD-grown monolayer graphene. Reproduced from [27] with permission; Copyright PNAS 2012.

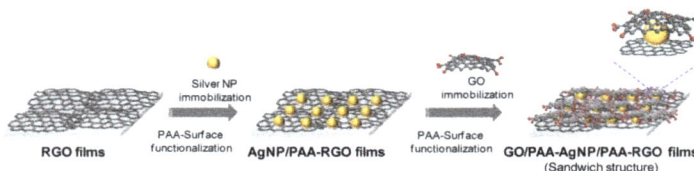

Figure 5. Fabrication Process of GO/PAA-AgNP/PAA-RGO films for application as SERS platform. Reproduced from [61] with permission; Copyright ACS 2012.

Besides ex-situ transferring of 2D materials onto metal surface, in-situ growing 2D materials, like graphene and MoS_2, on metal surface is another attractive approach for incorporating 2D materials with plasmonic structure. Liu et al. [65] developed a CVD process to grow graphene shell with controllable thickness on the surface of metal NPs. Figure 6 shows the fabrication process of graphene-encapsulated metal nanoparticles. In situ-grown 2D materials on metal surface do not require 2D materials transfer process and have little chance to have polymer residue left on the surface of 2D materials. CVD in-situ grown 2D materials is a promising method for conformally coating 2D materials on metallic surface. However, due to the high temperature of CVD process, pre-designed metallic nanostructure may change its morphology during high temperature process and lose the pre-designed high local EM field. Low temperature plasma-enhanced CVD method [66–68] could be a potential choice for in-situ growth of 2D materials on metal surface.

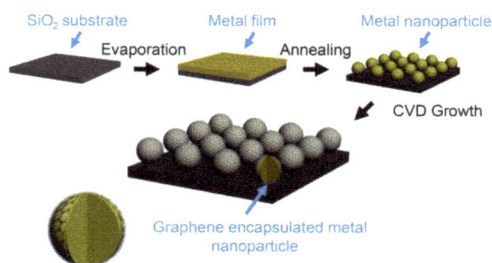

Figure 6. Production process for the Metal@Graphene to serve as a SERS-active substrate. Reproduced from [65] with permission; Copyright ACS 2014.

3.2. Sensitivity

Among various 2D materials, graphene is the most widely explored one for incorporation with plasmonic structure. Graphene/metal hybrid SERS platform shows superior SERS performance compared with bare metal SERS substrates. Because of chemical interaction between graphene and target molecules, certain SERS modes are enhanced or prohibited. Although the chemical enhancement factor of 2D materials is not as high as metallic nanostructure, several tens' of times of Raman signal enhancement could be essential when detecting molecules at single molecular level. Several times enhancement determines whether the Raman peaks can be seen or not. Comparisons of enhancement factors of different types of SERS sensors are summarized in Table 1. It can be seen that pristine 2D materials generally have lower enhancement factors than metallic SERS substrates or hybrid SERS substrates containing metal. However, a more meaningful comparison between different SERS platforms would require the same analytes and the same detection approach (laser wavelength, laser power, accumulation time, etc.).

Table 1. Comparisons of typical types of SERS substrates.

SERS Substrate Type	Substrate Materials	Enhancement Factor	Analytes	Ref.
Metal	Au nanotriangles	1.2×10^5	Benzenethiol	[69]
	Ag nanocubes	1.25×10^5	1,4-benzenedithiol	[70]
	Ag nanoparticles	10^{14}–10^{15}	Rhodamine 6G	[71]
Metal/inorganic hybrid structure	SiO$_2$-coated silver nanocubes	1.2×10^6	Rhodamine 6G	[72]
	Au nanoparticle-coated ZnO nanoneedles	1.2×10^7	Rhodamine 6G	[19]
	Au-coated ZnO nanorods	10^6	Rhodamine 6G	[73]
2D materials/Metal hybrid structure	Monolayer graphene-coated Au nanopyramids	10^{10}	Rhodamine 6G and lysozyme	[29]
	Monolayer graphene-coated Au nanovoids	10^3	Rhodamine 6G	[58]
	Few layer graphene-coated Au nanoparticles	9.2–19.4	Cobalt phthalocyanine	[65]
2D materials	Monolayer graphene	2–17	Phthalocyanine	[33]
	Monolayer MoS$_2$	5×10^4–3.8×10^5	4-mercaptopyridine	[31]
	h-BN	6.9–41	Copper phthalocyanine	[26]
	Monolayer WSe$_2$	0.18–4.7	Copper phthalocyanine	[32]
	Monolayer graphene on top of monolayer WS$_2$	3.8–78.2	Copper phthalocyanine	[32]

Wang et al. [29] developed a graphene/Au nanopyramid hybrid SERS platform, which shows single-molecule detection capability for analytes like R6G and lysozyme. In this study, SERS performance of graphene/Au nanopyramid hybrid substrate and bare Au nanopyramid substrate were compared. Table 2 [29] summarizes the average SERS signal enhancement of graphene/Au hybrid substrates over bare Au SERS substrate. As shown in Table 2, additional graphene layer contributes extra few times of Raman enhancement for R6G peak intensity compared with bare Au nanpyramid substrates, which demonstrates the synergic effect of electromagnetic and chemical enhancement from graphene. Even for molecules with small Raman cross-section, like dopamine and serotonin, graphene/Au hybrid platform can still achieve detection limit of 10^{-9} M in simulated body fluid [30]. With graphene/Au nanopyramid hybrid SERS substrates, serotonin molecule Raman peak hot spots and graphene peak hot spots actually coincide, as seen from the Raman intensity mapping of analytes peak with that of the graphene G peak (Figure 7). The results indicate that the intrinsic Raman signal of 2D materials in 2D materials/metal hybrid SERS platform can serve as a gauge of the near-field EM-field intensity to locate hot spots. This unique feature of hybrid platform offers an advantage for molecule detection at ultra-low concentrations. Actual hot spots of SERS substrates are rare and random, even for patterned nanostructure. For extremely diluted solution, the spatial coincidence of molecules and hotspots is rare, leading to long time of up to hours spent on searching for measurable signals. With 2D materials' intrinsic Raman peak intensity as a SERS enhancement factor marker, the hot spots of the 2D materials/metal hybrid SERS platform could be located in advance and speed up the later detection of target molecules. For 2D materials used in hybrid SERS platform with patterned metallic SERS nanostructures, graphene is the ideal choice, because graphene only has a few intrinsic

Raman peaks, and large-area, high quality graphene is easily achievable. In addition, monolayer graphene has only 2.3% absorption of the incident laser, and its plasmon resonance frequency is the tetra Hz regime. Therefore, it has little effect on the EM field of metallic SERS substrates.

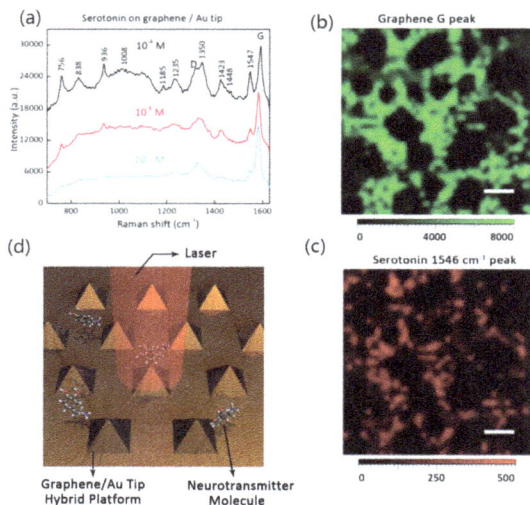

Figure 7. (**a**) Raman spectra of Serotonin molecules on graphene hybrid structure with 3 different concentrations (10^{-4}, 10^{-8}, and 10^{-10} M); (**b,c**) Raman intensity mapping of graphene G band (green) and Raman intensity mapping of serotonin peak at 1546 cm^{-1} (red) of the same area, scale bar: 10 μm; (**d**) Schematic of graphene/Au nanopyramid SERS substrate. Reproduced from [30] with permission; Copyright ACS 2015.

Table 2. Vibration mode dependent enhancement and assignment of Raman peaks in SERS spectra for R6G [29].

SERS Peaks (cm^{-1})	Peak Assignment	Average Enhancement ($I_{\text{Graphene/Au tip}}/I_{\text{Au tip}}$)
613	δ(C–C–C)ip	10
775	δ(C–H)op	5
1187	δ(C–H)ip	2
1311	ν(C–C)$^+$, ν(C–N)	6
1360	ν(C–C)$^+$, ν(C–N)	6
1506	ν(C–C)	4
1577	ν(C–O–C)	8

Note: $I_{\text{Graphene/Au tip}}$ is the R6G Raman intensity obtained on graphene/Au nanopyramid; $I_{\text{Au tip}}$ is the R6G Raman intensity obtained on bare Au nanopyramid.

Besides graphene, h-BN also served as coating layer on plasmonic structure for SERS application. Kim et al. [74] reported h-BN layer-wrapped Au nanoparticles as SERS substrate. h-BN coated Au SERS substrate can provide sensitive detection of aromatic hydrocarbon (PAC) molecules, such as B(α)P. PAC molecule Raman detection is very difficult using conventional metallic SERS, because the weak interaction between polycyclic aromatic hydrocarbon (PAC) molecules and the metal surface prohibits their adsorption on the metal surface. With h-BN wrapped Au SERS substrates, noticeable and characteristic bands of B(α)P can be detected (Figure 8a), because the π−π interaction between B(α)P and h-BN enlarges the surface adsorption coverage (Figure 8b).

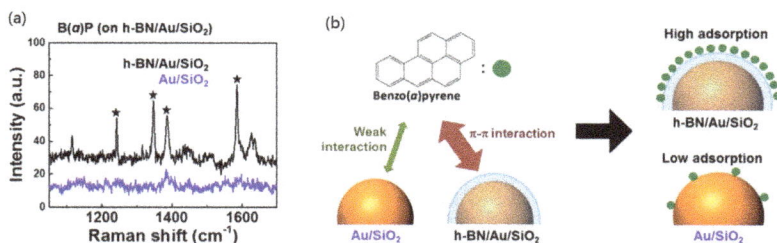

Figure 8. (**a**) SERS spectra benzo(α)pyrene on h-BN/Au/SiO$_2$ and Au/SiO$_2$ substrates; (**b**) schematic mechanism to explain SERS of benzo(α)pyrene on h-BN/Au/SiO$_2$ and Au/SiO$_2$ substrates. Reproduced from [74] with permission; Copyright ACS 2016.

3.3. Stability

Ag nano-structure is known to have excellent SERS performance with wider plasmonic spectral window than other metallic structure made of Au or Al. However, one major weakness of Ag nanostructure is that it is easily oxidized in ambient environment. The degradation of Ag will lower the SERS performance and cause uncertainty of analysis. In addition, photo induced damage on analyte molecules is a well know side effect of metallic SERS substrates. This section will discuss recent process of using 2D materials as shielding layer to protect SERS metal substrates from oxidation and protect analyte molecules from photo-induced damages.

2D materials, like graphene and h-BN, are able to protect metal to be oxidized [75,76]. This feature of 2D materials can also be used in SERS substrate development [65,77]. When single layer graphene combines with Ag nanostructure, the hybrid SERS platform provides both better SERS performance and excellent stability in a harsh environment (sulfur) and at high temperatures (300 °C) [78]. Liu et al. [77] combined CVD grown graphene with silver SERS substrates and demonstrated that with the graphene as protecting layer, the hybrid graphene/Ag SERS substrate could achieve large-area uniformity and long-term stability. Li et al. [79] compared the oxidation protection effect between CVD grown graphene and rGO coated Ag nanoparticles. They found out that CVD-grown monolayer graphene served as a better protecting layer than rGO to effectively suppress the oxidation of Ag nanoparticles. As seen from Figure 9, CVD-grown, graphene-coated Ag SERS substrate can provide stable R6G SERS signals up to 28 days with ambient aerobic exposure, while rapidly decreasing Raman signals are seen from rGO-coated and bare Ag nanoparticles. Worse performance of rGO-protected Ag nanoparticle is due to (1) the wide size distribution of rGO results an incoherent thin film and (2) the fact that cracks and holes on rGO film act as a channel to allow air reach Ag surface, leading to the oxidation of Ag nanoparticles.

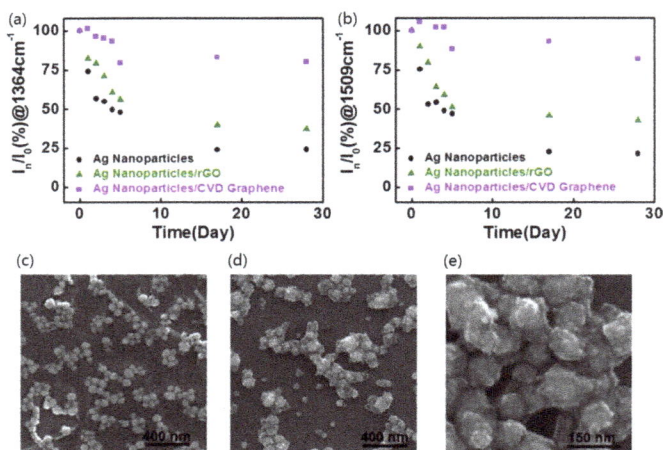

Figure 9. (**a**) The normalized intensity of the R6G Raman peak at 1364 cm^{-1} collected using, respectively, the unprotected (black), rGO-protected (green), and CVD graphene-protected (purple) Ag nanoparticles as substrates, versus the time of aerobic exposure; (**b**) the normalized intensity of the R6G Raman peak at 1509 cm^{-1} collected using, respectively, the unprotected (black), rGO-protected (green), and CVD graphene-protected (purple) Ag nanoparticles as substrates, versus the time of aerobic exposure; (**c**) SEM image of the CVD graphene-protected Ag nanoparticles after their 28-day use as the SERS substrate for the measurement of R6G; (**d**,**e**) SEM image of the unprotected Ag nanoparticles after their 28-day use as the SERS substrate for the measurement of R6G. Reproduced from [79] with permission; Copyright Elsevier 2013.

Another benefit to combine 2D materials with metallic nanostructures is that 2D materials can help protect molecules from photo-induced damage, such as photobleaching [27,65,80,81]. The photobleaching (or photodegradation) of the Raman anlaytes induced by the laser is a well-known side effect in SERS experiments, especially for dye molecules. When combining graphene with metallic nanostructure, the hybrid SERS platform is more stable against photo-induced damage with an even higher enhancement factor. Liu et al. [65] fabricated graphene-encapsulated metal nanoparticles for molecule detection and found out that AuNP/graphene hybrid substrate could significantly suppress photobleaching and fluorescence of cobalt phthalocyanine (CoPc) and R6G molecules. For instance, within the 160 s measurement period, the 1534 cm^{-1} peak intensity of CoPc molecules decreases dramatically for Au NPs, while the same peak intensity almost keeps constant for Au@Graphene, as shown in Figure 10a,b. Zhao et al. [81] also demonstrated that graphene can enhance the photostability of R6G molecules with graphene coated Ag SERS substrates during continuous light illumination. Enhanced photostability of molecules provided by graphene during SERS detection is attributed to $\pi-\pi$ interactions between graphene surface and molecules [65,81]. Molecule $\pi-\pi$ interaction with graphene allows the charge transfer between graphene and molecules, providing additional path for the molecules to relax from the excitation state to the ground state [82]. This process reduces the number of molecules at excitation states and thus decreases photobleaching rate. Similar protection effect can be achieved by using h-BN layer as well. Kim et al. [74] reported a h-BN film wrapped Au substrate showing extraordinary stability against photothermal and oxidative damages during laser excitation, as shown in Figure 10c,d. This outstanding stability against photothermal damage of h-BN wrapped Au SERS substrate is attributed to the ultrafast heat dissipation through the h-BN layer. With 2D materials as a shielding layer, hybrid SERS substrates will provide long-term stability.

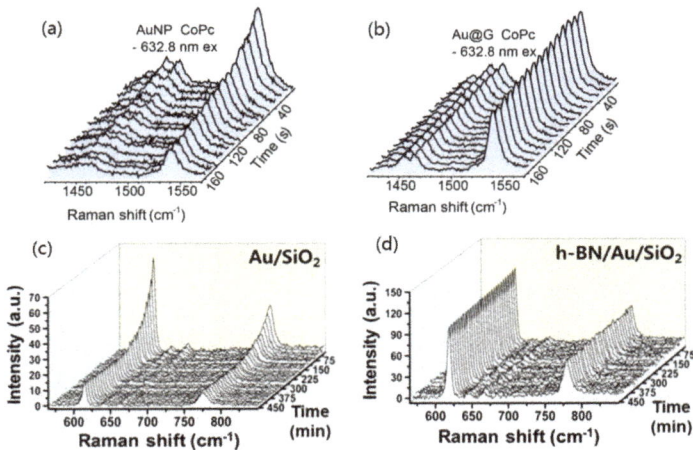

Figure 10. Stability of SERS signals of monolayer CoPc LB films on (**a**) Au and (**b**) Au@G. Reproduced from [65] with permission; Copyright ACS 2014. Photothermal and chemical stability of 3 L h-BN/Au/SiO2 substrate; SERS spectra of R6G on the Au/SiO2 substrate (**c**) without and (**d**) with h-BN protection at different time points (laser power = 0.1 mW, time interval = 15 min). Reproduced from [74] with permission; Copyright ACS 2016.

4. Conclusions and Perspective

In summary, 2D materials' Raman enhancement is due to the chemical enhancement mechanism, which differentiates them from metallic SERS substrates. Coating 2D materials on metallic SERS substrates introduces extra benefits over bare metal substrates. First, adding 2D materials can further increase the SERS enhancement factor due to the synergic effect of electromagnetic and chemical enhancement. Second, the atomic thin film of 2D materials can help map out the hot spots of the metallic nanostructure without affecting the local EM field of the metallic nanostructure underneath. For example, a Raman mapping of the graphene G peak over the hybrid SERS substrates could give the precise position of the hot spots. Finally, adding 2D materials as a shielding layer offers a chemically inert surface and helps to reduce the fluctuation of the SERS signal caused by the degradation of the metallic nano-structures, photobleaching, or metal-catalyzed site reactions, and thus improves the long-term stability and repeatability of the SERS analysis.

Conflicts of Interest: The authors declare no conflict of interest.

References

1. Pillai, I.C.; Li, S.; Romay, M.; Lam, L.; Lu, Y.; Huang, J.; Dillard, N.; Zemanova, M.; Rubbi, L.; Wang, Y. Cardiac fibroblasts adopt osteogenic fates and can be targeted to attenuate pathological heart calcification. *Cell Stem Cell* **2017**, *20*, 218–232. [CrossRef] [PubMed]

2. Yigit, M.V.; Medarova, Z. In vivo and ex vivo applications of gold nanoparticles for biomedical SERS imagingi. *Am. J. Nucl. Med. Mol. Imaging* **2012**, *2*, 232–241. [PubMed]

3. Qian, X.; Peng, X.-H.; Ansari, D.O.; Yin-Goen, Q.; Chen, G.Z.; Shin, D.M.; Yang, L.; Young, A.N.; Wang, M.D.; Nie, S. In vivo tumor targeting and spectroscopic detection with surface-enhanced Raman nanoparticle tags. *Nat. Biotechnol.* **2008**, *26*, 83–90. [CrossRef] [PubMed]

4. Motz, J.T.; Hunter, M.; Galindo, L.H.; Gardecki, J.A.; Kramer, J.R.; Dasari, R.R.; Feld, M.S. Optical fiber probe for biomedical Raman spectroscopy. *Appl. Opt.* **2004**, *43*, 542–554. [CrossRef] [PubMed]

5. Le Ru, E.C.; Etchegoin, P.G. Single-molecule surface-enhanced Raman spectroscopy. *Ann. Rev. Phys. Chem.* **2012**, *63*, 65–87. [CrossRef] [PubMed]

6. Luo, S.-C.; Sivashanmugan, K.; Liao, J.-D.; Yao, C.-K.; Peng, H.-C. Nanofabricated SERS-active substrates for single-molecule to virus detection in vitro: A review. *Biosens. Bioelectron.* **2014**, *61*, 232–240. [CrossRef] [PubMed]

7. Blackie, E.J.; Ru, E.C.L.; Etchegoin, P.G. Single-molecule surface-enhanced Raman spectroscopy of nonresonant molecules. *J. Am. Chem. Soc.* **2009**, *131*, 14466–14472. [CrossRef] [PubMed]

8. Dasary, S.S.; Singh, A.K.; Senapati, D.; Yu, H.; Ray, P.C. Gold nanoparticle based label-free SERS probe for ultrasensitive and selective detection of trinitrotoluene. *J. Am. Chem. Soc.* **2009**, *131*, 13806–13812. [CrossRef] [PubMed]

9. Lane, L.A.; Qian, X.; Nie, S. SERS nanoparticles in medicine: From label-free detection to spectroscopic tagging. *Chem. Rev.* **2015**, *115*, 10489–10529. [CrossRef] [PubMed]

10. Xia, M.; Zhang, P.; Qiao, K.; Bai, Y.; Xie, Y.-H. Coupling SPP with LSPR for enhanced field confinement: A simulation study. *J. Phys. Chem. C* **2015**, *120*, 527–533. [CrossRef]

11. Xia, M.; Zhang, P.; Leung, C.; Xie, Y.H. SERS optical fiber probe with plasmonic end-facet. *J. Raman Spectrosc.* **2017**, *48*, 211–216. [CrossRef]

12. Xia, M.; Qiao, K.; Cheng, Z.; Xie, Y.-H. Multiple layered metallic nanostructures for strong surface-enhanced Raman spectroscopy enhancement. *Appl. Phys. Express* **2016**, *9*, 065001. [CrossRef]

13. Yan, Z.; Xia, M.; Wang, P.; Zhang, P.; Liang, O.; Xie, Y.-H. Selective manipulation of molecules by electrostatic force and detection of single molecules in aqueous solution. *J. Phys. Chem. C* **2016**, *120*, 12765–12772. [CrossRef]

14. Bryche, J.-F.; Bélier, B.; Bartenlian, B.; Barbillon, G. Low-cost SERS substrates composed of hybrid nanoskittles for a highly sensitive sensing of chemical molecules. *Sens. Actuators B Chem.* **2017**, *239*, 795–799. [CrossRef]

15. Jimenez de Aberasturi, D.; Serrano-Montes, A.B.; Langer, J.; Henriksen-Lacey, M.; Parak, W.J.; Liz-Marzán, L.M. Surface enhanced Raman scattering encoded gold nanostars for multiplexed cell discrimination. *Chem. Mater.* **2016**, *28*, 6779–6790. [CrossRef]

16. Rodriguez-Fernandez, D.; Langer, J.; Henriksen-Lacey, M.; Liz-Marzán, L.M. Hybrid Au–SiO$_2$ core–satellite colloids as switchable SERS tags. *Chem. Mater.* **2015**, *27*, 2540–2545. [CrossRef]

17. Barbillon, G.; Sandana, V.E.; Humbert, C.; Bélier, B.; Rogers, D.J.; Teherani, F.H.; Bove, P.; McClintock, R.; Razeghi, M. Study of au coated ZnO nanoarrays for surface enhanced Raman scattering chemical sensing. *J. Mater. Chem. C* **2017**, *5*, 3528–3535. [CrossRef]

18. Cui, S.; Dai, Z.; Tian, Q.; Liu, J.; Xiao, X.; Jiang, C.; Wu, W.; Roy, V.A. Wetting properties and SERS applications of ZnO/Ag nanowire arrays patterned by a screen printing method. *J. Mater. Chem. C* **2016**, *4*, 6371–6379. [CrossRef]

19. Liu, L.; Yang, H.; Ren, X.; Tang, J.; Li, Y.; Zhang, X.; Cheng, Z. Au–ZnO hybrid nanoparticles exhibiting strong charge-transfer-induced SERS for recyclable SERS-active substrates. *Nanoscale* **2015**, *7*, 5147–5151. [CrossRef] [PubMed]

20. Magno, G.; Bélier, B.; Barbillon, G. Gold thickness impact on the enhancement of SERS detection in low-cost Au/Si nanosensors. *J. Mater. Sci.* **2017**, *52*, 13650–13656. [CrossRef]

21. Zhang, W.; Ma, R.; Chen, Q.; Xia, M.; Ng, J.; Wang, A.; Xie, Y.-H. The electro-mechanical responses of suspended graphene ribbons for electrostatic discharge applications. *Appl. Phys. Lett.* **2016**, *108*, 153103. [CrossRef]

22. Chen, Q.; Ma, R.; Lu, F.; Wang, C.; Liu, M.; Wang, A.; Zhang, W.; Xia, M.; Xie, Y.-H.; Cheng, Y. Systematic transient characterization of graphene interconnects for on-chip ESD protection. In Proceedings of the 2016 IEEE International Conference on Reliability Physics Symposium (IRPS), Pasadena, CA, USA, 17–21 April 2016; pp. 3B-6-1–3B-6-5.

23. Zhang, W.; Chen, Q.; Xia, M.; Ma, R.; Lu, F.; Wang, C.; Wang, A.; Xie, Y.-H. Tlp Evaluation of Esd Protection Capability of Graphene Micro-Ribbons for ICS. In Proceedings of the 2015 IEEE 11th International Conference on ASIC (ASICON), Chengdu, China, 3–6 November 2015; pp. 1–4.

24. Xia, F.; Mueller, T.; Lin, Y.-M.; Valdes-Garcia, A.; Avouris, P. Ultrafast graphene photodetector. *Nat. Nanotechnol.* **2009**, *4*, 839–843. [CrossRef] [PubMed]

25. Wang, X.; Wang, P.; Wang, J.; Hu, W.; Zhou, X.; Guo, N.; Huang, H.; Sun, S.; Shen, H.; Lin, T. Ultrasensitive and broadband MoS$_2$ photodetector driven by ferroelectrics. *Adv. Mater.* **2015**, *27*, 6575–6581. [CrossRef] [PubMed]

26. Ling, X.; Fang, W.; Lee, Y.-H.; Araujo, P.T.; Zhang, X.; Rodriguez-Nieva, J.F.; Lin, Y.; Zhang, J.; Kong, J.; Dresselhaus, M.S. Raman enhancement effect on two-dimensional layered materials: Graphene, h-BN and MoS$_2$. *Nano Lett.* **2014**, *14*, 3033–3040. [CrossRef] [PubMed]

27. Xu, W.; Ling, X.; Xiao, J.; Dresselhaus, M.S.; Kong, J.; Xu, H.; Liu, Z.; Zhang, J. Surface enhanced Raman spectroscopy on a flat graphene surface. *Proc. Natl. Acad. Sci.* **2012**, *109*, 9281–9286. [CrossRef] [PubMed]

28. Xu, W.; Mao, N.; Zhang, J. Graphene: A platform for surface-enhanced Raman spectroscopy. *Small* **2013**, *9*, 1206–1224. [CrossRef] [PubMed]

29. Wang, P.; Liang, O.; Zhang, W.; Schroeder, T.; Xie, Y.H. Ultra-sensitive graphene-plasmonic hybrid platform for label-free detection. *Adv. Mater.* **2013**, *25*, 4918–4924. [CrossRef] [PubMed]

30. Wang, P.; Xia, M.; Liang, O.; Sun, K.; Cipriano, A.F.; Schroeder, T.; Liu, H.; Xie, Y.-H. Label-free SERS selective detection of dopamine and serotonin using graphene-Au nanopyramid heterostructure. *Anal. Chem.* **2015**, *87*, 10255–10261. [CrossRef] [PubMed]

31. Muehlethaler, C.; Considine, C.R.; Menon, V.; Lin, W.-C.; Lee, Y.-H.; Lombardi, J.R. Ultrahigh Raman enhancement on monolayer MoS$_2$. *ACS Photonics* **2016**, *3*, 1164–1169. [CrossRef]

32. Tan, Y.; Ma, L.; Gao, Z.; Chen, M.; Chen, F. Two-dimensional heterostructure as a platform for surface-enhanced Raman scattering. *Nano Lett.* **2017**, *17*, 2621–2626. [CrossRef] [PubMed]

33. Ling, X.; Xie, L.; Fang, Y.; Xu, H.; Zhang, H.; Kong, J.; Dresselhaus, M.S.; Zhang, J.; Liu, Z. Can graphene be used as a substrate for Raman enhancement? *Nano Lett.* **2009**, *10*, 553–561. [CrossRef] [PubMed]

34. Xia, M. A review on applications of two-dimensional materials in surface-enhanced Raman spectroscopy. *Int. J. Spectrosc.* **2018**, *2018*, 4861472. [CrossRef]

35. Yan, Z.; Xia, M.; Zhang, P.; Xie, Y.H. Self-aligned trapping and detecting molecules using a plasmonic tweezer with an integrated electrostatic cell. *Adv. Op. Mater.* **2017**, *5*, 1600329. [CrossRef]

36. Yan, Z.; Liu, Z.; Xia, M.; Efimov, A.; Xie, Y.H. Broadband surface-enhanced coherent anti-stokes Raman spectroscopy with high spectral resolution. *J. Raman Spectrosc.* **2017**, *48*, 935–942. [CrossRef]

37. Ling, X.; Moura, L.; Pimenta, M.A.; Zhang, J. Charge-transfer mechanism in graphene-enhanced Raman scattering. *J. Phys. Chem. C* **2012**, *116*, 25112–25118. [CrossRef]

38. Ling, X.; Zhang, J. First-layer effect in graphene-enhanced Raman scattering. *Small* **2010**, *6*, 2020–2025. [CrossRef] [PubMed]

39. Liang, E.; Kiefer, W. Chemical effect of SERS with near-infrared excitation. *J. Raman Spectrosc.* **1996**, *27*, 879–886. [CrossRef]

40. Kneipp, K.; Wang, Y.; Kneipp, H.; Perelman, L.T.; Itzkan, I.; Dasari, R.R.; Feld, M.S. Single molecule detection using surface-enhanced Raman scattering (SERS). *Phys. Rev. Lett.* **1997**, *78*, 1667. [CrossRef]

41. Xu, H.; Aizpurua, J.; Käll, M.; Apell, P. Electromagnetic contributions to single-molecule sensitivity in surface-enhanced Raman scattering. *Phys. Rev. E* **2000**, *62*, 4318. [CrossRef]

42. Le Ru, E.C.; Etchegoin, P.G. *Principles of Surface-Enhanced Raman Spectroscopy: And Related Plasmonic Effects*; Elsevier: New York, NY, USA, 2008.

43. Moskovits, M. Surface-enhanced spectroscopy. *Rev. Mod. Phys.* **1985**, *57*, 783. [CrossRef]

44. Adrian, F.J. Charge transfer effects in surface-enhanced Raman scatteringa. *J. Chem. Phys.* **1982**, *77*, 5302–5314. [CrossRef]

45. Ling, X.; Wu, J.; Xu, W.; Zhang, J. Probing the effect of molecular orientation on the intensity of chemical enhancement using graphene-enhanced Raman spectroscopy. *Small* **2012**, *8*, 1365–1372. [CrossRef] [PubMed]

46. Dean, C.R.; Young, A.F.; Meric, I.; Lee, C.; Wang, L.; Sorgenfrei, S.; Watanabe, K.; Taniguchi, T.; Kim, P.; Shepard, K.L. Boron nitride substrates for high-quality graphene electronics. *Nat. Nanotechnol.* **2010**, *5*, 722. [CrossRef] [PubMed]

47. Splendiani, A.; Sun, L.; Zhang, Y.; Li, T.; Kim, J.; Chim, C.-Y.; Galli, G.; Wang, F. Emerging photoluminescence in monolayer MoS$_2$. *Nano Lett.* **2010**, *10*, 1271–1275. [CrossRef] [PubMed]

48. Liu, Y.; Yu, D.; Zeng, C.; Miao, Z.; Dai, L. Biocompatible graphene oxide-based glucose biosensors. *Langmuir* **2010**, *26*, 6158–6160. [CrossRef] [PubMed]

49. Li, N.; Zhang, Q.; Gao, S.; Song, Q.; Huang, R.; Wang, L.; Liu, L.; Dai, J.; Tang, M.; Cheng, G. Three-dimensional graphene foam as a biocompatible and conductive scaffold for neural stem cells. *Sci. Rep.* **2013**, *3*, 1604. [CrossRef] [PubMed]

50. Weng, Q.; Wang, B.; Wang, X.; Hanagata, N.; Li, X.; Liu, D.; Wang, X.; Jiang, X.; Bando, Y.; Golberg, D. Highly water-soluble, porous, and biocompatible boron nitrides for anticancer drug delivery. *ACS Nano* **2014**, *8*, 6123–6130. [CrossRef] [PubMed]

51. Zhu, Y.; Murali, S.; Cai, W.; Li, X.; Suk, J.W.; Potts, J.R.; Ruoff, R.S. Graphene and graphene oxide: Synthesis, properties, and applications. *Adv. Mater.* **2010**, *22*, 3906–3924. [CrossRef] [PubMed]

52. Li, X.; Zhu, H. Two-dimensional MoS_2: Properties, preparation, and applications. *J. Mater.* **2015**, *1*, 33–44. [CrossRef]

53. Zhang, K.; Feng, Y.; Wang, F.; Yang, Z.; Wang, J. Two dimensional hexagonal boron nitride (2D-hBN): Synthesis, properties and applications. *J. Mater. Chem. C* **2017**, *5*, 11992–12022. [CrossRef]

54. Fan, M.; Andrade, G.F.; Brolo, A.G. A review on the fabrication of substrates for surface enhanced Raman spectroscopy and their applications in analytical chemistry. *Anal. Chim. Acta* **2011**, *693*, 7–25. [CrossRef] [PubMed]

55. Brown, R.J.; Milton, M.J. Nanostructures and nanostructured substrates for surface—Enhanced Raman scattering (SERS). *J. Raman Spectrosc.* **2008**, *39*, 1313–1326. [CrossRef]

56. Ouyang, L.; Ren, W.; Zhu, L.; Irudayaraj, J. Prosperity to challenges: Recent approaches in SERS substrate fabrication. *Rev. Anal. Chem.* **2017**, *36*, 20160027. [CrossRef]

57. Xia, M.; Li, B.; Yin, K.; Capellini, G.; Niu, G.; Gong, Y.; Zhou, W.; Ajayan, P.M.; Xie, Y.-H. Spectroscopic signatures of AA′ and AB stacking of chemical vapor deposited bilayer MoS_2. *ACS Nano* **2015**, *9*, 12246–12254. [CrossRef] [PubMed]

58. Zhu, X.; Shi, L.; Schmidt, M.S.; Boisen, A.; Hansen, O.; Zi, J.; Xiao, S.; Mortensen, N.A. Enhanced light–matter interactions in graphene-covered gold nanovoid arrays. *Nano Lett.* **2013**, *13*, 4690–4696. [CrossRef] [PubMed]

59. Li, H.; Contryman, A.W.; Qian, X.; Ardakani, S.M.; Gong, Y.; Wang, X.; Weisse, J.M.; Lee, C.H.; Zhao, J.; Ajayan, P.M. Optoelectronic crystal of artificial atoms in strain-textured molybdenum disulphide. *Nat. Commun.* **2015**, *6*, 7381. [CrossRef] [PubMed]

60. Pirkle, A.; Chan, J.; Venugopal, A.; Hinojos, D.; Magnuson, C.; McDonnell, S.; Colombo, L.; Vogel, E.; Ruoff, R.; Wallace, R. The effect of chemical residues on the physical and electrical properties of chemical vapor deposited graphene transferred to SiO_2. *Appl. Phys. Lett.* **2011**, *99*, 122108. [CrossRef]

61. Kim, Y.-K.; Han, S.W.; Min, D.-H. Graphene oxide sheath on Ag nanoparticle/graphene hybrid films as an antioxidative coating and enhancer of surface-enhanced Raman scattering. *ACS Appl. Mater. Interfaces* **2012**, *4*, 6545–6551. [CrossRef] [PubMed]

62. Hu, Y.; Li, F.; Bai, X.; Li, D.; Hua, S.; Wang, K.; Niu, L. Label-free electrochemical impedance sensing of DNA hybridization based on functionalized graphene sheets. *Chem. Commun.* **2011**, *47*, 1743–1745. [CrossRef] [PubMed]

63. Georgakilas, V.; Otyepka, M.; Bourlinos, A.B.; Chandra, V.; Kim, N.; Kemp, K.C.; Hobza, P.; Zboril, R.; Kim, K.S. Functionalization of graphene: Covalent and non-covalent approaches, derivatives and applications. *Chem. Rev.* **2012**, *112*, 6156–6214. [CrossRef] [PubMed]

64. Shan, C.; Yang, H.; Han, D.; Zhang, Q.; Ivaska, A.; Niu, L. Water-soluble graphene covalently functionalized by biocompatible poly-L-lysine. *Langmuir* **2009**, *25*, 12030–12033. [CrossRef] [PubMed]

65. Liu, Y.; Hu, Y.; Zhang, J. Few-layer graphene-encapsulated metal nanoparticles for surface-enhanced Raman spectroscopy. *J. Phys. Chem. C* **2014**, *118*, 8993–8998. [CrossRef]

66. Wei, D.; Lu, Y.; Han, C.; Niu, T.; Chen, W.; Wee, A.T.S. Critical crystal growth of graphene on dielectric substrates at low temperature for electronic devices. *Angew. Chem. Int. Ed.* **2013**, *52*, 14121–14126. [CrossRef] [PubMed]

67. Qi, J.; Zheng, W.; Zheng, X.; Wang, X.; Tian, H. Relatively low temperature synthesis of graphene by radio frequency plasma enhanced chemical vapor deposition. *Appl. Surf. Sci.* **2011**, *257*, 6531–6534. [CrossRef]

68. Sun, J.; Chen, Y.; Cai, X.; Ma, B.; Chen, Z.; Priydarshi, M.K.; Chen, K.; Gao, T.; Song, X.; Ji, Q. Direct low-temperature synthesis of graphene on various glasses by plasma-enhanced chemical vapor deposition for versatile, cost-effective electrodes. *Nano Res.* **2015**, *8*, 3496–3504. [CrossRef]

69. Scarabelli, L.; Coronado-Puchau, M.; Giner-Casares, J.J.; Langer, J.; Liz-Marzán, L.M. Monodisperse gold nanotriangles: Size control, large-scale self-assembly, and performance in surface-enhanced Raman scattering. *ACS Nano* **2014**, *8*, 5833–5842. [CrossRef] [PubMed]

70. McLellan, J.M.; Li, Z.-Y.; Siekkinen, A.R.; Xia, Y. The SERS activity of a supported Ag nanocube strongly depends on its orientation relative to laser polarization. *Nano Lett.* **2007**, *7*, 1013–1017. [CrossRef] [PubMed]

71. Nie, S.; Emory, S.R. Probing single molecules and single nanoparticles by surface-enhanced Raman scattering. *Science* **1997**, *275*, 1102–1106. [CrossRef] [PubMed]

72. Kha, N.M.; Chen, C.-H.; Su, W.-N.; Rick, J.; Hwang, B.-J. Improved Raman and photoluminescence sensitivity achieved using bifunctional Ag@SiO$_2$ nanocubes. *Phys. Chem. Chem. Phys.* **2015**, *17*, 21226–21235. [CrossRef] [PubMed]

73. Sakano, T.; Tanaka, Y.; Nishimura, R.; Nedyalkov, N.N.; Atanasov, P.A.; Saiki, T.; Obara, M. Surface enhanced raman scattering properties using Au-coated ZnO nanorods grown by two-step, off-axis pulsed laser deposition. *J. Phys. D Appl. Phys.* **2008**, *41*, 235304. [CrossRef]

74. Kim, G.; Kim, M.; Hyun, C.; Hong, S.; Ma, K.Y.; Shin, H.S.; Lim, H. Hexagonal boron nitride/au substrate for manipulating surface plasmon and enhancing capability of surface-enhanced Raman spectroscopy. *ACS Nano* **2016**, *10*, 11156–11162. [CrossRef] [PubMed]

75. Liu, Z.; Gong, Y.; Zhou, W.; Ma, L.; Yu, J.; Idrobo, J.C.; Jung, J.; MacDonald, A.H.; Vajtai, R.; Lou, J. Ultrathin high-temperature oxidation-resistant coatings of hexagonal boron nitride. *Nat. Commun.* **2013**, *4*, 2541. [CrossRef] [PubMed]

76. Chen, S.; Brown, L.; Levendorf, M.; Cai, W.; Ju, S.-Y.; Edgeworth, J.; Li, X.; Magnuson, C.W.; Velamakanni, A.; Piner, R.D. Oxidation resistance of graphene-coated Cu and Cu/Ni alloy. *ACS Nano* **2011**, *5*, 1321–1327. [CrossRef] [PubMed]

77. Liu, X.; Wang, J.; Wu, Y.; Fan, T.; Xu, Y.; Tang, L.; Ying, Y. Compact shielding of graphene monolayer leads to extraordinary SERS-active substrate with large-area uniformity and long-term stability. *Sci. Rep.* **2015**, *5*, 17167. [CrossRef] [PubMed]

78. Hu, Y.; Kumar, P.; Xuan, Y.; Deng, B.; Qi, M.; Cheng, G.J. Controlled and stabilized light–matter interaction in graphene: Plasmonic film with large-scale 10-nm lithography. *Adv. Opt. Mater.* **2016**, *4*, 1811–1823. [CrossRef]

79. Li, X.; Li, J.; Zhou, X.; Ma, Y.; Zheng, Z.; Duan, X.; Qu, Y. Silver nanoparticles protected by monolayer graphene as a stabilized substrate for surface enhanced Raman spectroscopy. *Carbon* **2014**, *66*, 713–719. [CrossRef]

80. Du, Y.; Zhao, Y.; Qu, Y.; Chen, C.-H.; Chen, C.-M.; Chuang, C.-H.; Zhu, Y. Enhanced light–matter interaction of graphene–gold nanoparticle hybrid films for high-performance SERS detection. *J. Mater. Chem. C* **2014**, *2*, 4683–4691. [CrossRef]

81. Zhao, Y.; Xie, Y.; Bao, Z.; Tsang, Y.H.; Xie, L.; Chai, Y. Enhanced SERS stability of R6G molecules with monolayer graphene. *J. Phys. Chem. C* **2014**, *118*, 11827–11832. [CrossRef]

82. Xie, L.; Ling, X.; Fang, Y.; Zhang, J.; Liu, Z. Graphene as a substrate to suppress fluorescence in resonance Raman spectroscopy. *J. Am. Chem. Soc.* **2009**, *131*, 9890–9891. [CrossRef] [PubMed]

Article

Contamination-Free Graphene Transfer from Cu-Foil and Cu-Thin-Film/Sapphire

Jaeyeong Lee [1,2], Shinyoung Lee [1] and Hak Ki Yu [1,2,*]

[1] Department of Materials Science and Engineering, Ajou University, Suwon 16499, Korea; smuff20@ajou.ac.kr (J.L.); tonyfine92@gmail.com (S.L.)

[2] Department of Energy Systems Research, Ajou University, Suwon 16499, Korea

*   Correspondence: hakkiyu@ajou.ac.kr; Tel.: +82-31-219-1680; Fax: +82-31-219-1613

Academic Editors: Federico Cesano and Domenica Scarano
Received: 7 November 2017; Accepted: 27 November 2017; Published: 2 December 2017

Abstract: The separation of graphene grown on metallic catalyst by chemical vapor deposition (CVD) is essential for device applications. The transfer techniques of graphene from metallic catalyst to target substrate usually use the chemical etching method to dissolve the metallic catalyst. However, this causes not only high material cost but also environmental contamination in large-scale fabrication. We report a bubble transfer method to transfer graphene films to arbitrary substrate, which is nondestructive to both the graphene and the metallic catalyst. In addition, we report a type of metallic catalyst, which is 700 nm of Cu on sapphire substrate, which is hard enough to endure against any procedure in graphene growth and transfer. With the Cr adhesion layer between sapphire and Cu film, electrochemically delaminated graphene shows great quality during several growth cycles. The electrochemical bubble transfer method can offer high cost efficiency, little contamination and environmental advantages.

Keywords: graphene; bubble transfer; electrochemical delamination; Cu film; nondestructive; reusability

1. Introduction

Graphene, a two-dimensional honeycomb structure of carbon isotopes, is receiving worldwide attention due to its superior quality, such as high carrier mobility, excellent strength and high thermal conductivity [1,2]. Due to its high electrical conductivity, transparency, and flexibility, high-quality single-layer graphene with large area has received much attention in the industry [3–5]. However, the conventional method of transferring graphene essentially involves a chemical etching step to remove the metal substrate, which causes environmental pollution and damage to graphene, as well as an increase in the production cost [6]. In addition, it is not appropriate to transfer the graphene from some chemically inactive metal catalysts such as Ru, Au, and Pt because these types of metals are hard to remove completely and etching is costly. Recently, bubble-transfer method, which use electrochemical reaction to delaminate graphene from catalyst metal, was reported [7–10]. In this way, it is possible to save metal catalyst to synthesize graphene by re-using the catalyst. However, this method (conventionally using Cu foil of about 25 µm thickness) can cause other damages to the Cu foil during handling such as wrinkling, tearing, and vacuum retention in spin coating, resulting in defects in synthesized graphene.

Here we report the growth of graphene on three types of Cu catalysts. First, a conventional Cu foil was used [11,12]. Second, a Cu film deposited on a sapphire substrate without any other layer or processing was used [13,14]. Finally, we used a Cu film on a sapphire substrate with a Cr adhesive layer [15]. In these three samples, graphene was grown by CVD and transferred from Cu to target substrate by bubble transfer method. The sapphire substrate is robust enough to withstand mechanical damage, preventing Cu from tearing and wrinkling. In addition, the Cr layer between the sapphire and the Cu thin film improves the adhesion strength between both surfaces to prevent Cu film from peeling and tearing. Further, since Cr is more reactive than Cu, it also has an effect of preventing Cu from being oxidized. Therefore, this non-destructive method makes it possible to use the Cu film catalyst repeatedly without damages, and the synthesized graphene has almost the same quality as the original one.

2. Materials and Methods

2.1. Preparation

2.1.1. Catalyst Materials Preparation

(i) Commercial Cu foil (Alfa Aesar, Ward Hill, MA, USA, item no. 13382, 99.8% purity) which were first immersed for 1 min in 0.1 mole ammonium persulfate ((NH_4)$_2S_2O_8$, Sigma-Aldrich, St. Louis, MO, USA, item no. 248614, ACS reagent, 98.0%) solution to clean and etch the contaminants on the surface of the Cu foil; (ii) 700 nm Cu film on c-plane sapphire and (iii) 700 nm Cu film on adhesion layer (15 nm Cr) coated c-plane sapphire were prepared by using electron-beam evaporator (substrate temperature was room temp and chamber pressure was maintained at 0.05 mTorr during deposition).

2.1.2. CVD Process

Each catalyst is loaded into a quartz tube reaction chamber to perform graphene growth. The pressure in the chamber was pumped to 5 mTorr using a mechanical pump, then hydrogen (H_2) was injected into the chamber at 40 sccm. Then, the Cu film catalysts (Cu/c-sapphire and Cu/Cr/c-sapphire) temperature increased to 950 °C over 60 min. The reduction step of copper was done by flowing hydrogen gas at this stage. The pressure in the chamber is maintained at 500 mTorr. Then, methane (CH_4) gas was introduced into the chamber at 10 sccm for 10 min. Finally, furnace was rapidly cooled to room temperature to precipitate the graphene on the catalyst surface under a hydrogen gas flow of 40 sccm. For the Cu foil catalyst, the catalyst was heated to 1000 °C for 60 min and annealed during 30 min to increase grain size of Cu before methane injection, and the other step was the same with Cu film catalysts.

2.2. Bubble Transfer

After growth, the graphene sample was spin-coated with poly-methyl methacrylate (PMMA) at 1800 rpm for 40 s and dried in air for 10 min. 1 M NaOH aqueous solution was used in the electrolyte for constant current and voltage, 40 mA and 3 V. A platinum film was used as the electrode. The delaminated graphene/PMMA film was transferred to Si/SiO$_2$ substrate. The transferred graphene was heated at 180 °C for 30 min and then cleaned in acetone for 1 h. After that, it was cleaned in Isopropyl Alcohol (IPA) and Deionized (DI) water for 10 min each.

2.3. Analysis

Scanning electron microscope (SEM) images were taken using a JSM-6700 (JEOL, Tokyo, Japan) device at 5.0 kV. The Optical Microscope (OM) images were taken with a U-MSSP49 (Olympus, Tokyo, Japan) microscope. Raman spectra were obtained with a LabRAM HR Evolution—Nicolet iS50 (Jobin Yvon, Horiba, Kyoto, Japan) spectrometer under He–Ne 532 nm laser wavelength, 5 mm in diameter.

3. Results and Discussion

3.1. Bubble Transfer Method for Transferring Graphene

Figure 1 illustrates the limitations of the conventional graphene transfer method. The OM image in Figure 1a shows some voids and graphene flakes attached to the graphene. These voids are defects that occur during the transfer process. When there is an unattached region between the target substrate and the graphene, the graphene is teared and scratched during the process to dissolving PMMA [16–18]. For the graphene flakes, since the graphene synthesized on the copper foil grows on both sides of the foil, if it is not removed by the oxygen plasma, the residue as shown in Figure 1a remains beneath the surface. The oxygen plasma process is, however, a demanding process and can cause mechanical damage to the Cu foil during the handling process. The bubble transfer process allows clean graphene to be obtained without the oxygen plasma process. The bubble transfer method does not cause such damages to graphene, and only one side of the graphene can be clearly peeled off. Furthermore, it can also reduce the loss of Cu compared to the conventional chemical etching transfer method. Since the binding energy between graphene and Cu (0.033 eV per carbon atom) is relatively weak compared to the inter-planar bond strength of graphite, the graphene can be easily peeled off from Cu by the bubble transfer method [19–21]. Figure 1c shows a schematic view of bubble transfer. A copper foil/graphene/PMMA was used as the cathode, a platinum foil was used as the anode, and NaOH solution was used as the electrolyte. When direct current is applied, water is reduced on the surface of the Cu foil to generate hydrogen. When this hydrogen bubble occurs in large quantities between copper and graphene, the graphene/PMMA layer is separated from copper within seconds. PMMA coatings of sufficient thickness and concentration are needed, since thin PMMA/graphene film can be damaged by hydrogen bubbles or water [22,23]. The mass change of copper foil was measured during repeated CVD deposition and shown in Table 1. Compared to the original Cu foil which was 0.0235 g at first, it decreased by 0.0064 g for seven cycles. There was a decrease of about 0.001 g in one cycle. A slight reduction of about 4% per cycle is believed to have occurred during the repeated pre-etching and CVD processes as well as during the bubble transfer process.

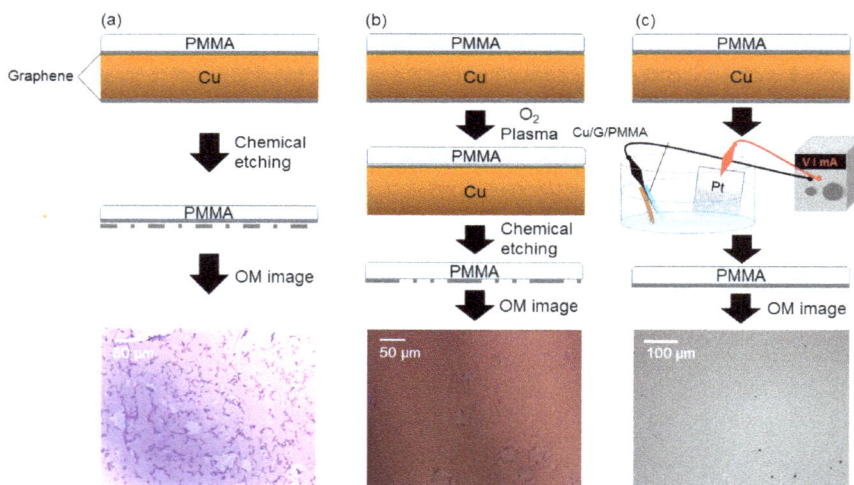

Figure 1. (**a**) Schematic image of chemical etching process of Cu without eliminating graphene below Cu; (**b**) Schematic image of chemical etching process of Cu after plasma etching for graphene below Cu; (**c**) Schematic image of bubble transfer method.

Table 1. Mass of the Cu during 7 cycles of CVD process.

No. of Cycle	1	2	3	4	5	6	7
Mass of Cu (g)	0.0235	0.0220	0.0210	0.0198	0.0191	0.0181	0.0171

In Figure 2, we analyzed the performance of graphene films made for seven cycles using a single Cu foil. As shown in Figure 2a, graphene films separated by the "bubble transfer" method have a considerably smaller amount of backside graphene residue which can trap the metal catalyst between transferred graphene film and backside graphene flake. Although not all metal residues can be eliminated with this method because the metal residues can exist not only at trapped surface but also at any other morphological feature such as grain boundaries, wrinkles, graphene adlayer, etc., it can be relatively reduced [24]. Additional research is needed to eliminate them. Figure 2b shows the Raman spectrum of seven graphene films with very high 2D/G peak ratio of about 3 to 10. This means that the synthesized graphene is close to a single layer. In addition, the very low intensity of the D-peak located at about 1350 cm^{-1} indicates that the qualities of these graphene films is very good. However, when Cu foil is used, the Cu foil may be damaged, torn or scratched by other transfer processes than bubble transfer (it even wrinkled easily with tweezers). Vacuum stamping, which is inevitable especially when spin coating PMMA, must be removed because it greatly degrades the performance of the graphene. So, we have grown a copper film on a sapphire substrate that is easy to handle in process and used it for graphene growth. C-plane sapphire (0001) was used due to low lattice mismatch with Cu (111) which is the most suitable catalytic surface for the hexagonal shaped graphene growth.

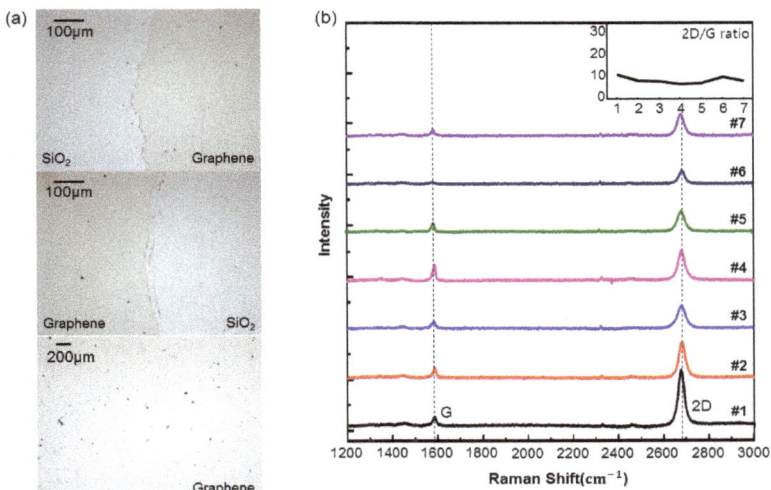

Figure 2. (**a**) OM images of graphene on Si/ SiO$_2$ transferred with bubble transfer method; (**b**) Raman spectrum of graphene made by a Cu foil used for seven cycles of graphene growth and transferred with bubble transfer method.

3.2. Effect of Cr Adhesive Layer

As shown in Figure 3, the adhesion between Cu and sapphire is poor, and Cu tends to peel off even once in bubble transfer. This is because bubble transfer is an electrochemical process that causes desorption on both sides of Cu. Figure 3b is an Optical Microscope (OM) image of a Cu surface after short-time electrolysis. As can be seen in the figure, the Cu layer is swollen, and when they are combined to some extent, they are separated from the sapphire substrate as shown in Figure 3a. To solve this problem, we added a Cr layer between Cu and sapphire. The Cr layer acts as an adhesive layer and strengthens the attraction between Cu and sapphire. Figure 3c shows the surface of a sample having a Cr layer after electrolysis, which is the same process as shown in Figure 3b. It is relatively more clean and well maintained than the former. We used XRD analysis to understand this effect (Figure 4). The status of Cu was compared by analyzing the XRD of Cu/Sapphire and Cu/Cr/Sapphire samples before and after 950 °C CVD graphene growth. The Cu (100)/Cu (200) intensity ratios of the two samples were 28 and 2.6 before the CVD process, respectively. Since the sapphire (0001) plane and the Cu (111) plane are hexagonal planes, the previous sample deposited directly on the sapphire is deposited in the (111) preferred orientation, unlike the latter where the Cr layer with the BCC structure is present [25,26]. It changes greatly after the CVD process. At high temperatures, the intensity of the (111) peak increases and the intensity of the (200) peak decreases due to Cu agglomeration.

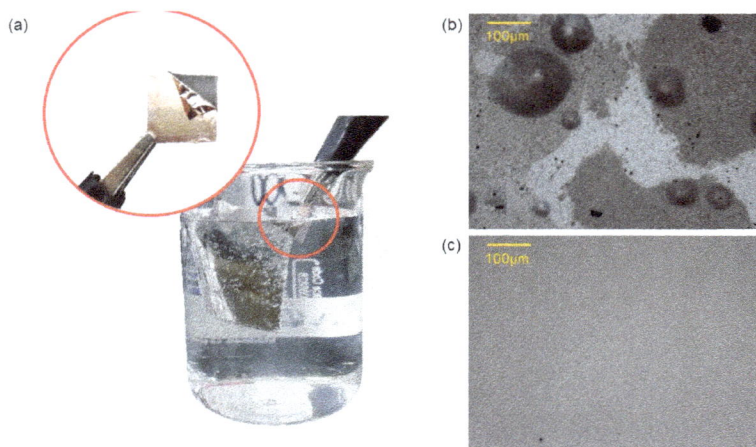

Figure 3. (**a**) Image of bubble transfer method for Sapphire/Cu and Cu which was detached from sapphire; (**b**,**c**) OM images of Sapphire/Cu and Sapphire/Cr/Cu after bubble-transfer.

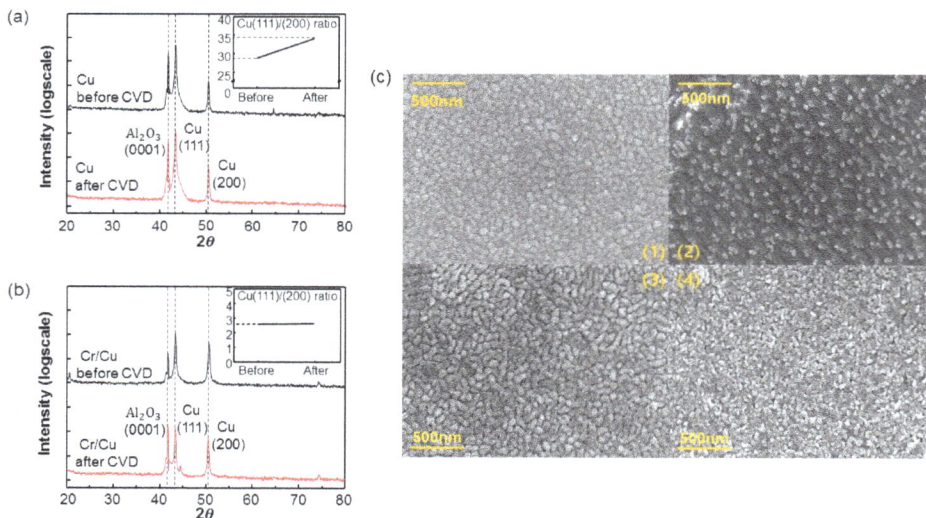

Figure 4. (**a**,**b**) XRD patterns and SEM images of Cu (**a**) and Cr/Cu (**b**) before and after CVD; (**c**) SEM images of Sapphire/Cu before CVD: (**1**), Sapphire/Cu after CVD (**2**), Sapphire/Cr/Cu before CVD (**3**) and Sapphire/Cr/Cu after CVD (**4**).

This means that Cu does not adhere well to the sapphire surface, and the performance of the Cu catalyst also deteriorates. The surface roughness of Cu is also expected to increase by agglomeration.

On the other hand, the sample containing the Cr layer showed no significant change in the intensity ratio before and after the CVD process. The Cu layer is stable on the Cr layer and it is very useful for repetitive graphene growth cycles. To ensure that Cu was agglomerated, we observed SEM images of the Cu surface of each sample [27]. As expected, there was no significant difference in the size of the surface particles in the sample containing the Cr layer, but samples without the Cr layer showed waves in the same direction due to the agglomeration of Cu. This is because the Cr located at the interface between Cu and Cr diffuses to the free surface and dielectric interface during CVD heating at 950 °C to make some Cu more adhesive [28]. It is interesting that small particles were generated from Cu without Cr after CVD process; they are expected to be generated with Cu oxide or contamination. The effect of Cr is also an evidence that it not only increases the adhesion but also prevents oxidation and contamination of Cu. We think that the oxygen in the sapphire substrate during the CVD process at high temperatures may have diffused and migrated to the Cu and Cr portions. In the absence of the Cr layer, diffused oxygen ions oxidize Cu, but in the case of the Cr layer, diffused oxygen combines with Cr to form Cr oxide, which acts as a passivation material [29,30]. This Cr oxide no longer causes oxygen to oxidize the Cu. The Cr-Cu phase diagram of Figure 5a shows that at 0.0169 wt % of Cr, which was calculated from our sample that has 700 nm Cu and 15 nm Cr, the two metals did not form solid solutions with each other at 950 °C.

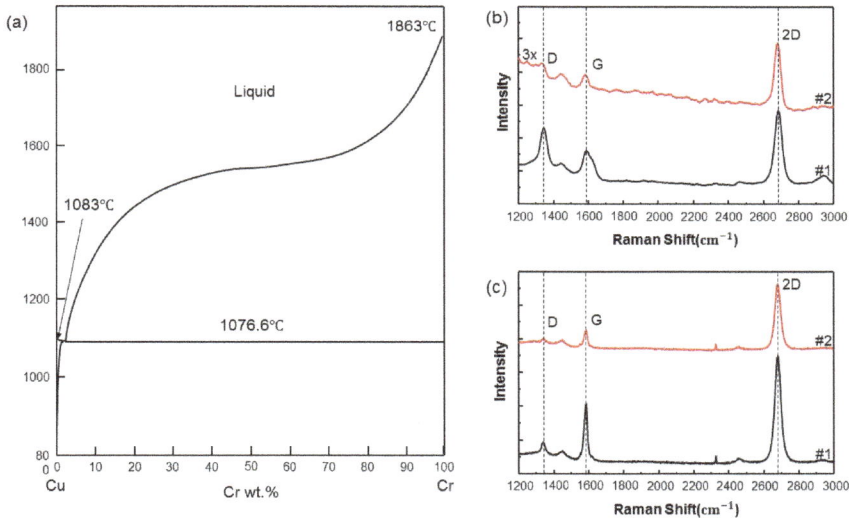

Figure 5. (**a**) Cu Cr phase diagram, Raman spectrum of graphene made at 950 °C (**b**), 970 °C (**c**).

3.3. Influence of Temperature

Experiments were performed at 930 °C, 950 °C, 970 °C and 1000 °C, respectively. However, the thermal energy at 930 °C was too low to dissolve carbon dissociated from methane into copper, so a complete graphene film could not be obtained. Too high a temperature at 1000 °C makes it difficult to synthesize complete graphene by evaporating too much copper during the CVD process. Raman spectra were observed after synthesis of graphene for two cycles at 950 and 970 °C, respectively. First, Figure 5b shows that the I_{2D}/I_G ratio of graphene synthesized at 950 °C is higher in the second synthesized graphene. D peak is also greater in the first synthesized graphene, which means that the second synthesized graphene has better quality than the former. This is because of the surface reconstruction of the Cu film at the CVD temperature which is close to the melting temperature of Cu [7]. This reconstruction makes the grain size of the Cu film large and uniform. The quality of the graphene, which is essentially influenced by the surface morphology of catalyst metal, can be increased by repeated use of the Cu film due to the surface reconstruction that occurred at the previous growth cycle. At 970 °C, as seen in Figure 5c, the performance of the second synthesized graphene is also better. Among them, graphene made at 970 °C has the best quality.

4. Conclusions

This study demonstrates the electrochemical exfoliation of graphene using Cu catalyst deposited on sapphire with Cr adhesion layer. Using Cr adhesive layer, we were able to produce high quality graphene films for repeated reuse and higher quality graphene at higher CVD temperatures. However, since too much Cu was blown off at 1000 °C, the optimum temperature for graphene generation was 970 °C. Other metals besides Cu can also be used as graphene growth catalysts such as nickel and cobalt. Bubble transfer can separate graphene from these metal catalysts without chemical etching [31–33]. Our next research project is to use the metals for graphene growth and bubble transfer.

Acknowledgments: This work was supported by the Ajou University research fund.

Author Contributions: Hak Ki Yu proposed the research topic; Jaeyeong Lee conceived and designed the experiments; Jaeyeong Lee and Shinyoung Lee performed the experiments; Jaeyeong Lee and Hak Ki Yu analyzed the data; Jaeyeong Lee and Hak Ki Yu wrote the paper.

Conflicts of Interest: The authors declare no conflict of interest.

References

1. Geim, A.K.; Novoselov, N.M.R. The rise of graphene. *Nat. Mater.* **2007**, *6*, 183–191. [CrossRef] [PubMed]
2. Neto, A.H.C.; Guinea, F.; Peres, N.M.R.; Novoselov, K.S.; Geim, A.K. The electronic properties of graphene. *Rev. Mod. Phys.* **2009**, *81*, 109–162. [CrossRef]
3. Huang, Y.; Dong, X.; Shi, Y.; Li, C.M.; Li, L.J.; Chen, P. Nanoelectronic biosensors based on CVD grown graphene. *Nanoscale* **2010**, *2*, 1485–1488. [CrossRef] [PubMed]
4. Cho, B.; Yoon, J.; Hahm, M.G.; Kim, D.H.; Kim, A.R.; Kahng, Y.H.; Park, S.W.; Lee, Y.J.; Park, S.G.; Kwon, J.D.; et al. Graphene-based gas sensor: Metal decoration effect and application to a flexible device. *J. Mater. Chem. C* **2014**, *2*, 5280–5285. [CrossRef]
5. Xia, F.; Mueller, T.; Lin, Y.; Valdes-Garcia, A.; Avouris, P. Ultrafast graphene photodetector. *Nat. Nanotechnol.* **2009**, *4*, 839–843. [CrossRef] [PubMed]
6. Liang, X.; Sperling, B.A.; Calizo, I.; Cheng, G.; Hacker, C.A.; Zhang, Q.; Obeng, Y.; Yan, K.; Peng, H.; Li, Q.; et al. Toward clean and crackless transfer of graphene. *ACS Nano* **2011**, *5*, 9144–9153. [CrossRef] [PubMed]
7. Wang, Y.; Zheng, Y.; Xu, X.; Dubuisson, E.; Bao, Q.; Lu, J.; Loh, K. Electrochemical delamination of CVD-grown graphene film: Toward the recyclable use of Cu catalyst. *ACS Nano* **2011**, *5*, 9927–9933. [CrossRef] [PubMed]
8. Gao, L.; Ren, W.; Xu, H.; Jin, L.; Wang, Z.; Ma, T.; Ma, L.; Zhang, Z.; Fu, Q.; Peng, L.; et al. Repeated growth and bubbling transfer of graphene with millimetre-size single-crystal grains using platinum. *Nat. Commun.* **2012**, *3*, 699. [CrossRef] [PubMed]
9. Cherian, C.T.; Giustiniano, F.; Martin-Fernandez, I.; Andersen, H.; Balakrishnan, J.; Ozyilmaz, B. 'Bubble-free' electrochemical delamination of CVD graphene films. *Small* **2015**, *11*, 189–194. [CrossRef] [PubMed]
10. Liu, L.; Liu, X.; Zhan, Z.; Guo, W.; Xu, C.; Deng, J.; Chakarov, D.; Hyldgaard, P.; Schroder, E.; Yurgens, A.; et al. A mechanism for highly efficient electrochemical bubbling delamination of CVD-grown from metal substrates. *Adv. Mater. Interfaces* **2016**, *3*, 1500492. [CrossRef]
11. Li, X.; Cai, W.; An, J.; Kim, S.; Nah, J.; Yang, D.; Piner, R.; Velamakanni, A.; Jung, I.; Tutuc, E.; et al. Large-area synthesis of high-quality and uniform graphene films on Cu foils. *Science* **2009**, *324*, 1312–1314. [CrossRef] [PubMed]
12. Li, X.; Colombo, L.; Ruoff, R. Synthesis of graphene films on Cu foils by chemical vapor deposition. *Adv. Mater.* **2016**, *28*, 6247–6252. [CrossRef] [PubMed]
13. Reddy, K.; Gledhill, A.; Chen, C.; Drexler, J.; Padture, N. High quality, transferrable graphene grown on single crystal Cu(111) thin films on basal-plane sapphire. *Appl. Phys. Lett.* **2011**, *98*, 113117. [CrossRef]

14. Baoshan, H.; Hiroki, A.; Yoshito, I.; Kenji, K.; Masaharu, T.; Eisuke, M.; Kazushi, S.; Noriaki, M.; Ken-ichi, I.; Seigi, M. Epitaxial growth of large-area single-layer graphene over Cu(111)/sapphire by atmospheric pressure CVD. *Carbon* **2012**, *50*, 57–65. [CrossRef]

15. Chan, C.; Chang, C.; Farrell, C.; Schrott, A. Adhesion studies of Cu-Cr alloys on Al₂O₃. *Appl. Phys. Lett.* **1993**, *62*, 654–656. [CrossRef]

16. Li, X.; Zhu, Y.; Cai, W.; Borysiak, M.; Han, B.; Chen, D.; Piner, R.D.; Colomba, L.; Ruoff, R.S. Transfer of large-area graphene films for high-performance transparent conductive electrodes. *Nano Lett.* **2009**, *9*, 4359–4363. [CrossRef] [PubMed]

17. Suk, J.W.; Kitt, A.; Magnuson, C.W.; Hao, Y.; Ahmed, S.; An, J.; Swan, A.K.; Goldberg, B.B.; Ruoff, R.S. Transfer of CVD-grown monolayer graphene onto arbitrary substrates. *ACS Nano* **2011**, *5*, 6916–6924. [CrossRef] [PubMed]

18. Wang, X.; Tao, L.; Hao, Y.; Liu, Z.; Chou, H.; Kholmanov, I.; Chen, S.; Tan, C.; Jayant, N.; Yu, Q.; et al. Direct delamination of graphene for high-performance plastic electronics. *Small* **2014**, *10*, 694–698. [CrossRef] [PubMed]

19. Giovannetti, G.; Khomyakov, P.A.; Brocks, G.; Karpan, V.M.; van den Brink, J.; Kelly, P.J. Doping graphene with metal contacts. *Phys. Rev. Lett.* **2008**, *101*, 4–7. [CrossRef] [PubMed]

20. Schabel, M.C.; Martins, J.L. Energetics of interplanar binding in graphite. *Phys. Rev. B* **1992**, *46*, 7185–7188. [CrossRef]

21. Taleb, A.A.; Yu, H.K.; Anemone, G.; Farias, D.; Wodtke, A.M. Helium diffraction and acoustic phonons of graphene grown on copper foil. *Carbon* **2015**, *95*, 731–737. [CrossRef]

22. Zhan, Z.; Sun, J.; Liu, L.; Wang, E.; Cao, Y.; Lindvall, N.; Skoblin, G.; Yurgens, A. Pore-free bubbling delamination of chemical vapor deposited graphene from copper foils. *J. Mater. Chem. C* **2015**, *3*, 8634–8641. [CrossRef]

23. De La Rosa, C.J.L.; Sun, J.; Lindvall, N.; Cole, M.T.; Nam, Y.; Löffler, M.; Olsson, E.; Teo, K.B.K.; Yurgens, A. Frame assisted H₂O electrolysis induced H₂ bubbling transfer of large area graphene grown by chemical vapor deposition on Cu. *Appl. Phys. Lett.* **2013**, *102*, 2011–2015. [CrossRef]

24. Lupina, G.; Kitzmann, J.; Costina, I.; Lukosius, M.; Wenger, C.; Wolff, A.; Vaziri, S.; Östling, M.; Pasternak, I.; Krajewska, A.; et al. Residual metallic contamination of transferred chemical vapor deposited graphene. *ACS Nano* **2015**, *9*, 4776–4785. [CrossRef] [PubMed]

25. Zhao, G.L.; Smith, J.R.; Raynolds, J.; Srolovitz, D.J. First-principles study of the α-Al₂O₃(0001)/Cu(111) interface. *Interface Sci.* **1996**, *3*, 289–302. [CrossRef]

26. Hall, M.G.; Aaronson, H.I.; Kinsma, K.R. The structure of nearly coherent fcc: Bcc boundaries in a CuCr alloy. *Surf. Sci.* **1972**, *31*, 257–274. [CrossRef]

27. Son, J.H.; Yu, H.K.; Song, Y.H.; Kim, B.J.; Lee, J.-L. Design of epitaxially strained Ag film for durable Ag-based contact to p-type GaN. *Cryst. Growth Des.* **2011**, *11*, 4943–4949. [CrossRef]

28. Russeel, S.W.; Rafalski, S.A.; Spreitzer, R.L.; Li, J.; Moinpour, M.; Moghadam, F.; Alford, T.L. Enhanced adhesion of copper to dielectrics via titanium and chromium additions and sacrificial reactions. *Thin Solid Film* **1995**, *262*, 154–167. [CrossRef]

29. Fu, G.Y.; Niu, Y.; Gesmundo, F. Microstructural effects on the high temperature oxidation of two-phase Cu-Cr alloys in 1atm O₂. *Corros. Sci.* **2003**, *45*, 559–574. [CrossRef]

30. Sirringhaus, H.; Theiss, S.D.; Kahn, A.; Wagner, S. Self-passivated copper gates for amorphous silicon thin-film transistors. *IEEE Electron Device Lett.* **1997**, *18*, 388–390. [CrossRef]

31. Gong, Y.; Zhang, X.; Liu, G.; Wu, L.; Geng, X.; Long, M.; Cao, X.; Guo, Y.; Li, W.; Sun, M.; et al. Layer-controlled and wafer-scale synthesis of uniform and high-quality graphene films on a polycrystalline nickel catalyst. *Adv. Funct. Mater.* **2012**, *22*, 3153–3159. [CrossRef]

32. Weatherup, R.S.; Bayer, B.C.; Blume, R.; Ducati, C.; Baehtz, C.; Schlogl, R.; Hofmann, S. In situ characterization of alloy catalysts for low-temperature graphene growth. *Nano Lett.* **2011**, *11*, 4154–4160. [CrossRef] [PubMed]

33. Losurdo, M.; Giangregorio, M.M.; Capezzuto, P.; Bruno, G. Graphene CVD growth on copper and nickel: Role of hydrogen in kinetics and structure. *Phys. Chem. Chem. Phys.* **2011**, *13*, 20836. [CrossRef] [PubMed]

coatings

MDPI

Article

Effect of Electrode Coating with Graphene Suspension on Power Generation of Microbial Fuel Cells

Hung-Yin Tsai [1], Wei-Hsuan Hsu [2,*] and Yi-Jhu Liao [2]

[1] Department of Power Mechanical Engineering, National Tsing Hua University, Hsinchu 30013, Taiwan; hytsai@pme.nthu.edu.tw

[2] Department of Mechanical Engineering, National United University, Miaoli 36003, Taiwan; M0211012@nuu.edu.tw

*   Correspondence: whhsu@nuu.edu.tw; Tel.: +886-37-382318

Received: 1 April 2018; Accepted: 18 June 2018; Published: 10 July 2018

Abstract: Microbial fuel cells (MFCs), which can generate low-pollution power through microbial decomposition, are a potentially vital technology with applications in environmental protection and energy recovery. The electrode materials used in MFCs are crucial determinants of their capacity to generate electricity. In this study, we proposed an electrode surface modification method to enhance the bacterial adhesion and increase the power generation in MFCs. Graphene suspension (GS) is selected as modifying reagent, and thin films of graphene are fabricated on an electrode substrate by spin-coating. Application of this method makes it easy to control the thickness of graphene film. Moreover, the method has the advantage of low cost and large-area fabrication. To understand the practicality of the method, the effects of the number of coating layers and drying temperature of the graphene films on the MFCs' performance levels are investigated. The results indicate that when the baking temperature is increased from 150 to 325 °C, MFC power generation can increase approximately 4.5 times. Besides, the maximum power density of MFCs equipped with a four-layer graphene anode is approximately four times that of MFCs equipped with a two-layer graphene anode. An increase in baking temperature or number of coating layers of graphene films enhances the performance of MFC power generation. The reason can be attributed to the graphene purity and amount of graphene adhering to the surface of electrode.

Keywords: microbial fuel cells; stainless steel mesh electrode; graphene; graphene suspension; air-cathode

1. Introduction

In the last decade, renewable energy sources that emit little pollution have been extensively studied due to shortages of energy and the rise of environmental awareness. Microbial fuel cells (MFCs) are one solution to this problem. MFCs utilize microorganisms as catalysts to break the chemical bonds of organic compounds and harvest electrical energy [1]. In the 1910s, the concept of applying microorganisms as catalysts in fuel cell systems was first explored [2]. The technology of fuel cell systems was not yet mature, and systems could produce only weak electricity. Thus, MFC technology did not receive any public attention at that time. Breakthroughs in fuel cell technology and energy crises have led to renewed interest in the development of MFCs.

The advantage of microbial fuel cells is that they can treat wastewater and produce electricity at the same time. Different levels of wastewater have been treated using MFCs technology, such as distillery wastewater [3], industrial wastewater [4], livestock wastewater [5], and domestic wastewater [6,7]. In these studies, bioelectricity generation is mainly achieved using natural microflora. The type of natural microflora affects the efficiency of wastewater treatment and electricity generation. Thus, some studies use cultivated bacteria to reduce biological variability as a source of noise. *Escherichia coli* (*E. coil*) is a commonly used cultivated bacteria for the study of electrode design [8–10].

The high cost of fabrication and low power output may be major obstacles toward the commercialization of MFCs. Some factors that affect the cost of an MFC include the type of reactor, the membrane separator, the electrode materials, and catalyst materials [11–13]. Typical MFC configurations include double-chamber MFCs [8], flat-plate MFCs [6], and single-chamber MFCs [7,9,10,14]. Single-chamber MFC configurations have the advantage of higher power generation and smaller volume than double-chamber MFCs [15]. Moreover, single-chamber MFC configurations can reduce cost by eliminating proton exchange membranes (PEMs) [14]. Thus, single-chamber MFCs have potential for commercial development.

The power output of an MFC is dependent on operational conditions and several factors, such as microbial inoculation, electrode materials, ionic concentration, catalyst, internal resistance, and electrode spacing [16–19]. Among these factors, electrode materials exert critical effects on the active surface utilization. Metal electrodes, such as stainless steel and titanium, have become a research focus due to their high electrical conductivity, which can effectively collect electrons and reduce ohmic loss [9,20]. Moreover, in MFC systems, electrons are generated by electrochemically active microorganisms at the interfaces between anodic surfaces and microbes [21,22]. Microorganisms grown as a biofilm on the anode surface are crucial for the performance of MFCs. Biofilm is like an electron acceptor of anode, which directly affects substrate metabolism and electronic collection capability [23,24]. For facilitating bacterial attachment and subsequent biofilm formation, three-dimensional metal, for example stainless steel mesh (SSM) [9,18–22], has been used as an electrode substrate. Electrode surface coatings with nanomaterials have become a reliable and effective approach for enhancing the power output and reliability of MFCs.

Graphene has been intensively studied as a possible electrode material for MFCs [8–10,25] due to its properties, namely high electrical conductivity, surface area, and stability. Anodes modified with graphene can improve the electrode surface area, the adhesion of bacteria, and the efficiency of electron transfer [8]. For cathode modification, graphene can be used as a catalyst for oxygen reduction reactions [9,10].

Most published studies have used the soaking method and chemical vapor deposition (CVD) to coat graphene onto the surfaces of electrodes. Although the soaking method has the advantage of being a simple process, it is difficult to control the coating thickness of the graphene. Moreover, during the soaking process, both waste and pollution from the graphene solution are major issues. CVD easily controls the deposition thickness of graphene and has superior deposition uniformity. However, expensive equipment and relatively long deposition times are not conducive to commercial development. Due to this coating process problem, the current study proposed a coating method for graphene deposition on electrode surfaces. Graphene suspension (GS) is spin-coated on SSM electrodes to obtain large and uniform graphene films. The effects of the number of coating layers and drying temperature of the graphene films on the MFCs' performance levels are the primary research content required to understand the practicality of this method. The power densities of the MFCs under different electrode spin coating conditions were evaluated as the performance indices.

2. Materials and Methods

2.1. MFC System and Preparation of Electrodes

The single-chamber MFC that we studied is shown in Figure 1. Figure 1a shows a schematic of the air-cathode MFC, which is cylindrical with a diameter of 40 mm, a length of 60 mm, and a total reactor volume of approximately 75 mL. Figure 1b shows a cross-section of the MFC. The cathode electrode and PEM (Nafion 117, Dupont Co., Wilmington, DE, USA) were fixed on the air-side, whereas the anode electrode was fixed on the opposite side of the cylindrical chamber. The base material of the electrodes was 304 SSM (E Shie Zong Co., Ltd., Taiwan) with an average diameter of approximately 30 μm (400 mesh). The modified material for SSM electrode is GS (S-WB30, Enerage Inc., Yilan, Taiwan). S-WB30 was prepared using 3 wt % of graphene in water as a solvent. The specific surface area of graphene was larger than 15 m^2/g. The average sheet thickness of graphene sheets was over 5 nm, and the lateral size of graphene sheets was approximately 20 μm. To coat the GS on the surfaces of the SSM electrodes, we proposed a surface modification method. For anodic modification, the SSM was washed by acetone, alcohol, and deionized water and then dried by a hot plate at 150 °C. After the cleaning process, the SSM was spin-coated (1500 rmp, 30 s) with different number of GS layers by using a spin coater (C-SP-M1-S, Power Assist Instrument Scientific Corp., Taiwan) and dried by a hot plate at 195 to 325 °C. After drying, graphene adsorbed onto the SSM was stable, and there was evidence that the carbon metal bonds between the carbon material and steel could be formed through the heating process [26]. The result was not only affected by the heating temperature but also by other factors, such as material geometry size. The interaction between nanosize objects and flat substrates has been reported as size-dependent [27]. According to the coating results in this study, the most obvious disadvantage of using more than four layers of GS was an overly thick graphene layer. To avoid shedding a thick graphene layer from the SSM surface, this study focused on the effect of SSM electrodes with 2–4 GS layers on MFC performance. After the coating process, the sheet resistivity of the modified electrode was measured by the four-point probe technique (QT-50, Quatek Co., Ltd., Taiwan). The aim of this study was to investigate the effect of the number of coating layers and drying temperature of anodic modification on the performance of MFC.

For cathodic modification, we used the same procedure as the anode electrode to clean and dry the SSM cathode. After the cleaning process, the SSM was spin-coated (1500 rmp, 30 s) with one layer of GS and dried by a hot plate at 325 °C. Then, the SSM cathode was spin-coated with poly-tetrafluoroethylene (PTFE, 60 wt % dispersion in H$_2$O, Sigma-Aldrich, St. Louis, MO, USA) for waterproofing. The coating speed was maintained at 1000 rpm and a bake temperature of 340 °C. According to the testing result, the coating process of PTFE had to be repeated four times to provide excellent waterproofing for the cathode.

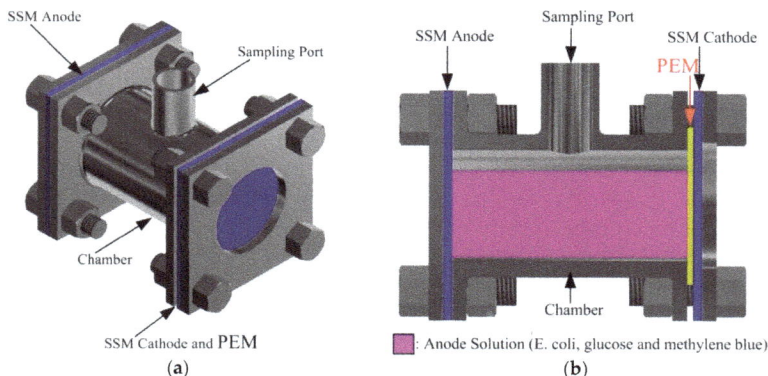

Figure 1. Schematic (**a**) and cross-section (**b**) of the single-chamber microbial fuel cells (MFCs) used in the experiment.

2.2. Microorganisms and Anode Solution

A single bacterium, *Escherichia coli* (*E. coli*) HB101, was used to convert energy to reduce the experimental variability and precisely estimate the effect of electrode modification of the MFCs' performance. To facilitate the comparison of data, 9-h cultures of HB101 cell and methylene blue were used in the MFC system. Glucose was used as fuel, and the anode solution was prepared by dispersing 0.1 g of methylene blue powder and 6.9 g of glucose powder in 102.5 g of *E. coli* solution.

2.3. Measurements and Analyses

The morphology of graphene was characterized with scanning electron microscopy (SEM, JSM-6500, Jeol Co., Tokyo, Japan) at 15 kV. To investigate the effect of the baking temperature of the electrode on graphene performance, thermogravimetric analysis (TGA, TGA 2950, Du Pont Instruments, Wilmington, DE, USA), and energy-dispersive X-ray spectroscopy (EDX, JSM-5600, Jeol Co., Tokyo, Japan) were employed to analyze the weight change of the GS and the surface composition of graphene at different temperatures.

The electrochemical experiments were carried out on an electrochemical workstation (AUT85126, Metrohm, Herisau, Switzerland). A three-electrode arrangement was used, consisting of an Ag/AgCl reference electrode, a working electrode, and a platinum counter electrode. Polarization and power density curves were used to evaluate the performance of the MFCs. Thus, a linear sweep voltammetry method was applied to evaluate the overpotential and current production rates at different applied voltages, and the measurements were performed at a controlled temperature of 30 °C. Power density P (W m^{-2}) was calculated according to the equation $P = IV/A$, in which I (A) is the current, V (V) is the voltage, and A (m^2) is the projected cross-sectional area of the anode.

3. Results and Discussion

3.1. GS Coated Electrodes

Figure 2 displays the surface of mesh electrodes after two layer (Figure 2a,d) and four layers (Figure 2e,h) of spin coating with graphene suspension (GS). Increasing the number of coating layers increased the amount of graphene adhered to the electrodes as well as the uniformity of the graphene. Figure 3 depicts the resistance of electrodes with 1–4 layers of GS coating. The electrode coated with four layers of GS demonstrated the optimal resistance of 62 MΩ cm^{-1}. The result indicated that the increased amount of adhered graphene and the improved uniformity of GS coating due to an increase in GS coating layers enhanced the electrode conductivity.

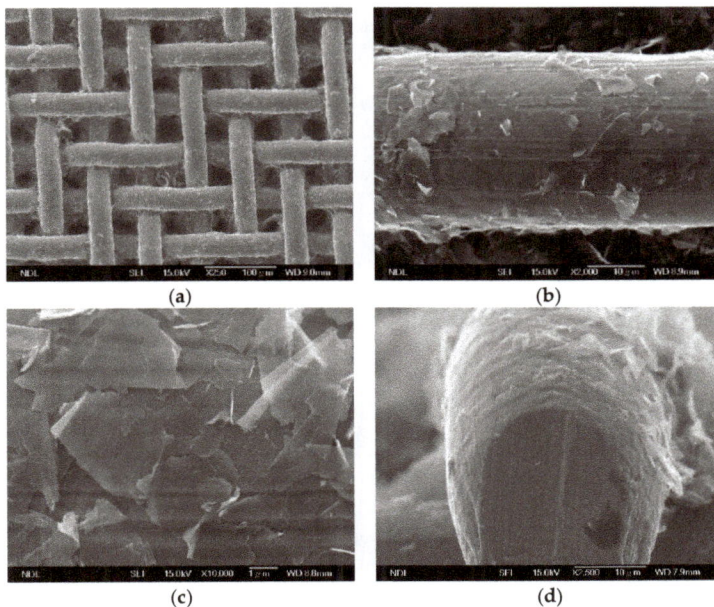

(a)

(b)

(c)

(d)

Figure 2. *Cont.*

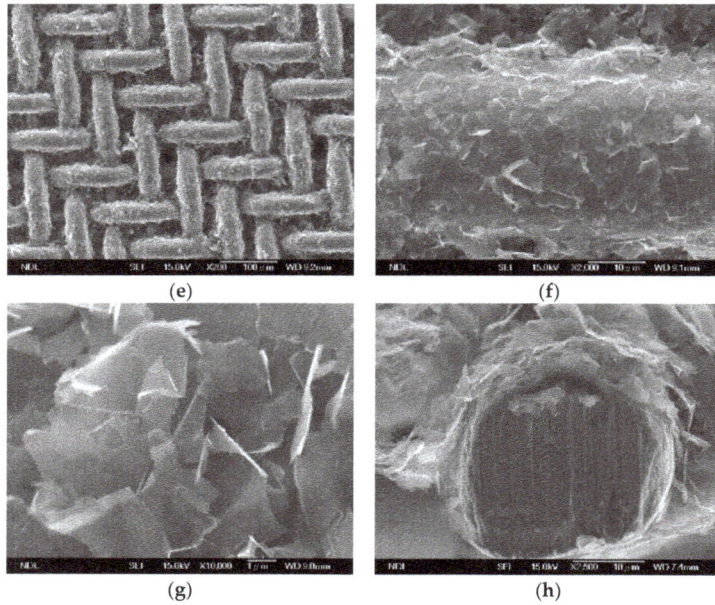

Figure 2. Surface of stainless steel mesh electrodes after one layer (**a–d**) and four layers (**e–h**) of spin coating with graphene suspension (GS).

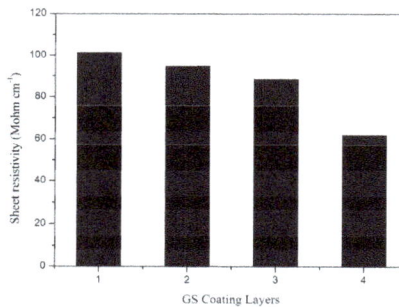

Figure 3. Effect of GS coating on electrode resistance

3.2. Effect of Electrode Baking Temperature on MFC Performance

Commercial GS contains dispersants. After GS coating, the electrodes were baked at various temperatures to examine the effect of residual dispersants on microbial fuel cell (MFC) performance. Figure 4 provides the linear scan voltammetry (LSV) spectra of MFC electrodes baked at various temperatures. Specifically, baking temperatures of 195, 250, and 325 °C led to maximal power densities of 0.27, 0.62, and 1.57 mW m^{-2}, respectively (Figure 3b,c and Figure 4a). After the MFC was assembled using electrodes spin coated with GS and dried at 195 °C, the cell performance notably improved 6 h after the assembly, but the maximum power density remained lower than those of the MFCs with electrodes dried at 250 and 325 °C. The aforementioned phenomenon did not occur when drying temperatures were 250 and 325 °C. At these two temperatures, the MFC power output first increased with time and then gradually decreased, finally reaching stability. Therefore, this study inferred that increasing the electrode drying temperature enhanced the decomposition of dispersants in the GS coating and improved the power density and cell stability.

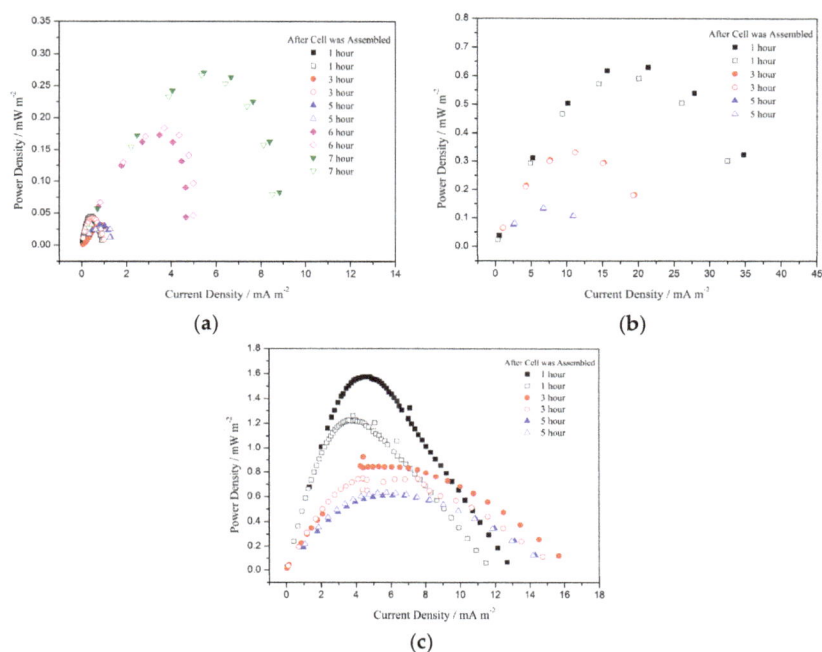

Figure 4. Current–power density curves of MFCs measured between 1 and 7 h after the cells were assembled; the electrodes were dried at temperatures of (**a**) 195, (**b**) 250, and (**c**) 325 °C.

To further investigate the effect of baking temperature on graphene characteristics, this study explored the variations in the weight of GS coating and graphene surface composition under different temperatures using thermogravimetric analysis (TGA) and energy-dispersive X-ray spectroscopy (EDX). Figure 5a depicts the change in the weight percentage of the GS from room temperature to 400 °C. Because when the temperature exceeded 100 °C, the change in GS weight percentage was subtle, the weight change between 125 and 400 °C is separately displayed in Figure 5b. In the heating process, when the temperature was increased from room temperature to approximately 125 °C, the water content in GS evaporated rapidly, which led to a rapid decrease of the weight GS. When the temperature was higher than 125 °C, the weight of GS started to slowly decrease. The water in GS was expected to have evaporated completely when the temperature exceeded 130 °C, but when the temperature was increased from 130 to 300 °C, the variation in the weight of the GS was approximately 20%. This variation indicated that a considerable amount of dispersant was retained at 130 °C, and the small weight loss might be attributable to the decomposition of solid composition in GS. Thus, the result indicated that increasing the baking temperature of coated electrodes facilitated graphene purification.

EDX (JSM-5600, Jeol Co., Tokyo, Japan) analysis and Raman spectroscopy (IHR550, Horiba Jobin Yvon, Kyoto, Japan) were used to evaluate the quality and layer stacking of graphene baked at different temperatures, as illustrated in Figure 6. EDX analysis of Figure 6a indicates that the percentage of the atomic concentration of carbon increased from 85% to 95% as the baking temperature increased from 195 to 325 °C. The proportion of carbon increased as the baking temperature increased. This result was illustrated through Raman spectroscopy and can be seen in Figure 6b. Raman spectroscopy is a reliable tool for evaluating the quality and layer stacking of graphene. The intensity ratio of D peak to G peak, $I(D)/I(G)$, provides information regarding the level of disorder in terms of covalent modification of graphene. When the baking temperature increased from 195 to 325 °C, the intensity ratio of the D peak to G peak decreased from 0.13 to 0.07. An increase in the proportion of carbon on the surface of graphene is a result of the decrease in covalent bond characters. The results mean that the probability of adsorption between graphene surface and with other substances is reduced as the baking temperature increased. Furthermore, the intensity ratio of 2D peak to G peak, $I(2D)/I(G)$, provides information regarding the layer stacking of graphene. In our study, the intensity ratio of 2D peak to G peak was similar and approximately 0.4. The graphene we used was multilayer, and the structure was not significantly affected by the baking temperature. According to the aforementioned results, an increase in the baking temperature of GS improved the quality of graphene and improved MFC performance. In the next two sections, the observation of biofilm formation and effect of the number of GS coating layers on MFC performance are explored using coated electrodes baked at 325 °C.

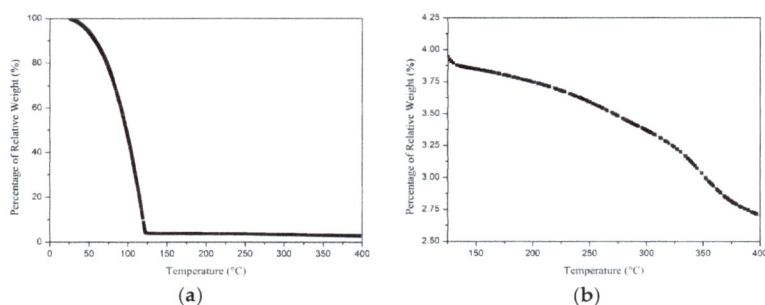

Figure 5. Thermogravimetric analysis (TGA) results: (**a**) variation in weight percentage of GS coating from room temperature to 400 °C; (**b**) subtle variation in weight percentage of GS coating at 125–400 °C.

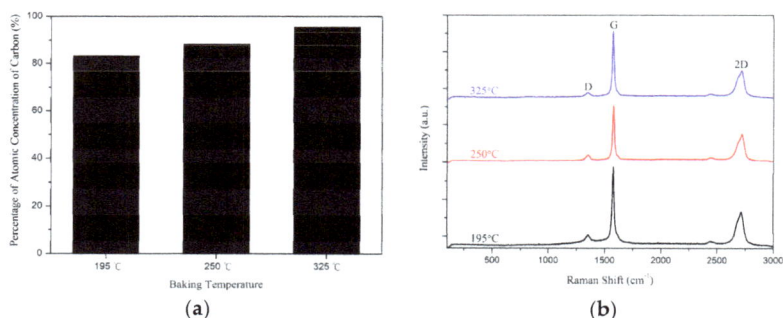

Figure 6. (**a**) X-ray spectroscopy (EDX) analysis and (**b**) Raman spectroscopy were employed to evaluate the quality and layer stacking of graphene baked at different temperatures.

3.3. Biofilm Morphology

The biofilm morphology was characterized by SEM. Figure 7 shows the surface morphology of the anode coated with two layers of GS at various times after the cells were assembled. When the cell operating time was less than 3 h, the number of microorganisms attached to the anode surface increased as the operating time increased, as exhibited in Figure 7a,b. However, when the cell operating time was greater than 5 h, the number of microorganisms attached to the anode surface only slightly increased, as shown in Figure 7c–e. The biofilm formation time in the MFC system was approximately 3 h.

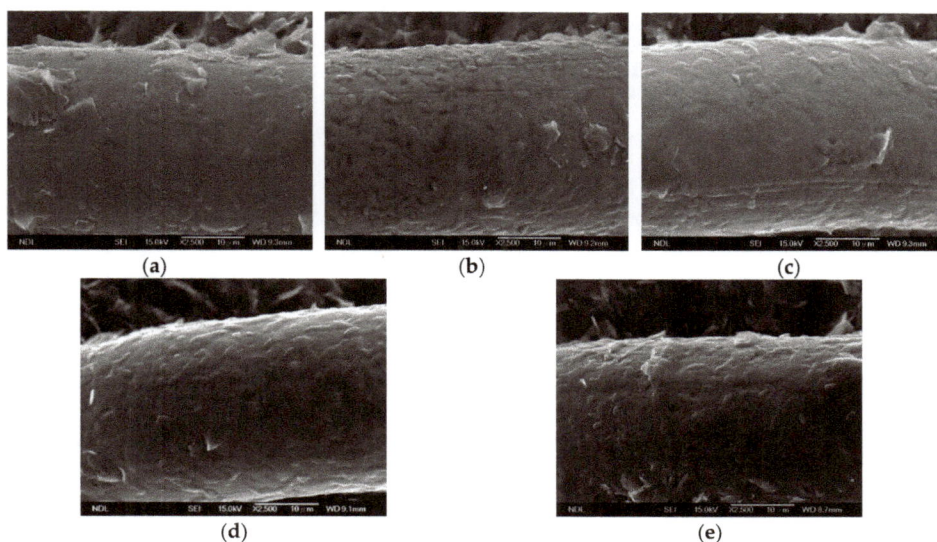

Figure 7. Surface morphology of the anode at (**a**) 1 h, (**b**) 2 h, (**c**) 3 h, (**d**) 5 h, and (**e**) 10 h after the cells were assembled.

3.4. Effect of the Number of GS Coated Layers on MFC Performance

To investigate the effect of the number of GS coated layers on MFC performance, anodes were spin coated with 2–4 layers of GS and baked at 325 °C. The substrates for cathodes and anodes were identical; one layer of GS and polytetrafluoroethylene was spin coated on the surface as the catalyst and waterproofing layers, respectively. Notably, to ensure all the data were acquired when the cells were at a stable state, the output voltage variation over time was assessed after all the cells were assembled with the experimental electrodes. Figure 8 shows the voltage output from the SSM anode coated with two layers of GS a long period after the cells were assembled. The red line in the Figure 7 illustrates the result of adding *E. coli* to the anode tank, and the black line shows the result of not adding *E. coli*. We considered the MFC system to have reached a stable value when the output signal variation was less than 5% of its steady-state value. When the SSM anode was coated with two layers of GS, the MFC system was able to reach a stable state after the cell was assembled for 218 min, and the steady state value of the voltage output was approximately 40 mV. When *E. coli* was not added to the anode tank, the voltage output was approximately 0.7 mV, equivalent to noise. Thus, in the studied MFC system, the conversion of energy was through *E. coli* and occurred in a favorable experimental environment. In addition, the anodes coated with three and four layers of GS required 225 and 232 min for MFCs to reach a stable state, respectively. Increasing the thickness of graphene coated on electrode substrates only slightly affected MFCs' stabilizing time. Moreover, the time required for the system to enter a steady state was similar to the biofilm formation time.

To ensure experimental precision, based on the evaluation results of MFCs, LSV measurement was conducted 5 h after the cells were assembled. In Figure 9, 2GL, 3GL, and 4GL are the LSV results of cell anodes coated with two, three, and four layers of GS, respectively. The open circuit voltage (OCV) of 2GL, 3GL, and 4GL were 0.23, 0.43, and 0.42 V, respectively, whereas the maximum power density of 2GL, 3GL, and 4GL were 0.44, 0.61, and 1.77 mW m^{-2}, respectively. The maximum power density of the cell equipped with a four-layer graphene anode was approximately four times that of the cell equipped with a two-layer graphene anode. Increasing the number of spin coated layers improved the amount of graphene adhering to the surface of the stainless-steel mesh. Thus, the specific surface area and conductivity of the electrodes were effectively increased, resulting in improved MFC performance.

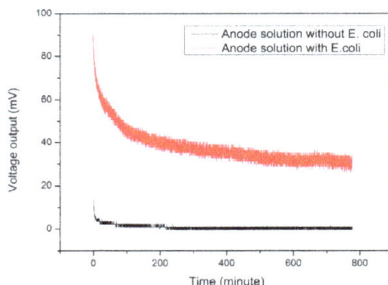

Figure 8. Effect of *E. coli* on the voltage output of the MFC system.

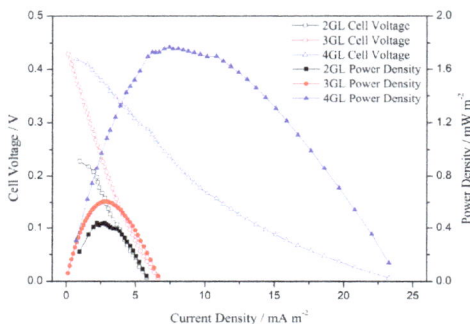

Figure 9. Linear scan voltammetry (LSV) results of anodes coated with two to four layers of GS.

4. Conclusions

This paper proposes an innovative method for stainless steel mesh modification: using GS spin coating technology to control the number of coating layers and the adhesion amount of graphene on electrodes. Because commercial GS contained dispersants, a heating process was required for the thermodecomposition of the dispersants. To explore the effect of residual dispersants on graphene quality, this study investigated the variation in the weight of the GS coating, the surface compositions and defect of graphene under various temperatures through TGA, EDX analysis, and Raman spectroscopy. The results showed that when the baking temperature was higher than 325 °C, the weight change of the graphene coating was less than 1%, and the carbon proportion on the surface of graphene exceeded 95%. Moreover, an increase in the baking temperature resulted a decrease in covalent bond characters of graphene surface. This indicated that this temperature could effectively decompose the dispersants in graphene and reduce the amount of dispersant residue on the graphene surface, resulting in an increase in graphene activity.

After deciding the required drying temperature of commercial GS coating on stainless steel mesh electrodes, this study explored the effect of the number of GS spin coating layers on MFC performance. By observing the coating using scanning electron microscopy and measuring the resistance using four-point probes, the increase in the number of coating layers was proved to be effective in increasing the amount of graphene adhered to the electrodes and the conductivity of the electrodes. Furthermore, the results of LSV measurements proved that the maximum power density increased four-fold (from 0.44 to 1.77 mW m^{-2}) when GS spin coating increased from two to four layers.

The proposed electrode modification method is simple in its process and substantially reduces graphene consumption in the coating process. Therefore, this method effectively reduces the cost of electrode modification. By controlling the number of GS spin coating layers and the drying temperature of electrodes, this study revealed the effect of the modification parameters on MFC performance and verified the feasibility of the proposed method. This method can be used to facilitate MFC development.

Author Contributions: Conceptualization, H.-Y.T. and W.-H.H.; Methodology, H.-Y.T. and W.-H.H.; Software, W.-H.H. and Y.-J.L.; Validation, W.-H.H. and Y.-J.L.; Formal Analysis, W.-H.H. and Y.-J.L.; Investigation, H.-Y.T. and W.-H.H.; Resources, H.-Y.T. and W.-H.H.; Data Curation, H.-Y.T. and W.-H.H.; Writing-Original Draft Preparation, W.-H.H.; Writing-Review & Editing, H.-Y.T.; Visualization, H.-Y.T. and W.-H.H.; Supervision, W.-H.H.; Project Administration, W.-H.H.; Funding Acquisition, H.-Y.T. and W.-H.H.

Funding: This research received no external funding.

Acknowledgments: The authors are grateful to National Nano Device Laboratories for helping SEM imaging.

Conflicts of Interest: The authors declare no conflict of interest.

References

1. Lovley, D.R. Bug juice: Harvesting electricity with microorganisms. *Nat. Rev. Microbiol.* **2006**, *4*, 497–508. [CrossRef] [PubMed]
2. Potter, M.C. Electrical effects accompanying the decomposition of organic compounds. *Proc. R. Soc. Lond. B* **1911**, *84*, 260–276. [CrossRef]
3. Lin, C.W.; Wu, C.H.; Huang, W.T.; Tsai, S.L. Evaluation of different cell-immobilization strategies for simultaneous distillery wastewater treatment and electricity generation in microbial fuel cells. *Fuel* **2015**, *144*, 1–8. [CrossRef]
4. Abbasi, U.; Jin, W.; Pervez, A.; Bhatti, Z.A.; Tariq, M.; Shaheen, S.; Iqbal, A.; Mahmood, Q. Anaerobic microbial fuel cell treating combined industrial wastewater: Correlation of electricity generation with pollutants. *Bioresour. Technol.* **2016**, *200*, 1–7. [CrossRef] [PubMed]
5. Doherty, L.; Zhao, Y.; Zhao, X.; Wang, W. Nutrient and organics removal from swine slurry with simultaneous electricity generation in an alum sludge-based constructed wetland incorporating microbial fuel cell technology. *Chem. Eng. J.* **2015**, *266*, 74–81. [CrossRef]
6. Min, B.; Logan, B.E. Continuous electricity generation from domestic wastewater and organic substrates in a flat plate microbial fuel cell. *Environ. Sci. Technol.* **2004**, *38*, 5809–5814. [CrossRef] [PubMed]
7. Liu, H.; Ramnarayanan, R.; Logan, B.E. Production of electricity during wastewater treatment using a single chamber microbial fuel cell. *Environ. Sci. Technol.* **2004**, *38*, 2281–2285. [CrossRef] [PubMed]
8. Zhang, Y.; Mo, G.; Li, X.; Zhang, W.; Zhang, J.; Ye, J.; Huang, X.; Yu, C. A graphene modified anode to improve the performance of microbial fuel cells. *J. Power Sources* **2011**, *196*, 5402–5407. [CrossRef]
9. Hsu, W.H.; Tsai, H.Y.; Huang, Y.C. Characteristics of carbon nanotubes/graphene coatings on stainless steel meshes used as electrodes for air-cathode microbial fuel cells. *J. Nanomater.* **2017**, *2017*, 9875301. [CrossRef]
10. Tsai, H.Y.; Hsu, W.H.; Huang, Y.C. Characteristics of carbon nanotube/graphene on carbon cloth as electrode for air-cathode microbial fuel cell. *J. Nanomater.* **2015**, *2015*, 686891. [CrossRef]
11. Liu, J.; Feng, Y.; Wang, X.; Shi, X.; Yang, Q.; Lee, H.; Zhang, Z.; Ren, N. The use of double-sided cloth without diffusion layers as air-cathode in microbial fuel cells. *J. Power Sources* **2011**, *196*, 8409–8412. [CrossRef]
12. Noori, Md.T.; Mukherjee, C.K.; Ghangrekar, M.M. Enhancing performance of microbial fuel cell by using graphene supported V_2O_5-nanorod catalytic cathode. *Electrochim. Acta* **2017**, *228*, 513–521. [CrossRef]
13. Noori, Md.T.; Ghangrekar, M.M.; Mukherjee, C.K. V_2O_5 microflower decorated cathode for enhancing power generation in air-cathode microbial fuel cell treating fish market wastewater. *Int. J. Hydrogen Energy* **2016**, *41*, 3638–3645. [CrossRef]
14. Liu, H.; Logan, B.E. Electricity generation using an air-cathode single chamber microbial fuel cell in the presence and absence of a proton exchange membrane. *Environ. Sci. Technol.* **2004**, *38*, 4040–4046. [CrossRef] [PubMed]
15. Yang, S.; Jia, B.; Liu, H. Effects of the Pt loading side and cathode-biofilm on the performance of a membrane-less and single-chamber microbial fuel cell. *Bioresour. Technol.* **2009**, *100*, 1197–1202. [CrossRef] [PubMed]

16. Cercado-Quezada, B.; Delia, M.L.; Bergel, A. Treatment of dairy wastes with a microbial anode formed from garden compost. *J. Appl. Electrochem.* **2010**, *40*, 225–232. [CrossRef]

17. Menicucci, J.; Beyenal, H.; Marsili, E.; Veluchamy, R.A.; Demir, G.; Lewandowski, Z. Procedure for determining maximum sustainable power generated by microbial fuel cells. *Environ. Sci. Technol.* **2006**, *40*, 1062–1068. [CrossRef] [PubMed]

18. Logan, B.E.; Murano, C.; Scott, K.; Gray, N.D.; Head, I.M. Electricity generation from cysteine in a microbial fuel cell. *Water Res.* **2005**, *39*, 942–952. [CrossRef] [PubMed]

19. Cheng, S.; Liu, H.; Logan, B.E. Increased power generation in a continuous flow MFC with advective flow through the porous anode and reduced electrode spacing. *Environ. Sci. Technol.* **2006**, *40*, 2426–2432. [CrossRef] [PubMed]

20. Wei, J.; Liang, P.; Huang, X. Recent progress in electrodes for microbial fuel cells. *Bioresour. Technol.* **2011**, *102*, 9335–9344. [CrossRef] [PubMed]

21. Zhi, W.; Ge, Z.; He, Z.; Zhang, H. Methods for understanding microbial community structures and functions in microbial fuel cells. *Bioresour. Technol.* **2014**, *171*, 461–468. [CrossRef] [PubMed]

22. Malvankar, N.S.; Lovley, D.R. Microbial nanowires: A new paradigm for biological electron transfer and bioelectronics. *ChemSusChem* **2012**, *5*, 1039–1046. [CrossRef] [PubMed]

23. Schröder, U. Anodic electron transfer mechanisms in microbial fuel cells and their energy efficiency. *Phys. Chem. Chem. Phys.* **2007**, *9*, 2619–2629. [CrossRef] [PubMed]

24. Cheng, S.; Logan, B.E. Ammonia treatment of carbon cloth anodes to enhance power generation of microbial fuel cells. *Electrochem. Commun.* **2007**, *9*, 492–496. [CrossRef]

25. Hou, J.; Liu, Z.; Li, Y.; Yang, S.; Zhou, Y. A comparative study of graphene-coated stainless steel fiber felt and carbon cloth as anodes in MFCs. *Bioprocess Biosyst. Eng.* **2015**, *38*, 881–888. [CrossRef] [PubMed]

26. Craig, S.; Harding, G.L.; Payling, R. Auger lineshape analysis of carbon bonding in sputtered metal-carbon thin films. *Surf. Sci.* **1983**, *124*, 591–601. [CrossRef]

27. Gao, H.J.; Wang, X.; Yao, H.M.; Gorb, S.; Arzt, E. Mechanics of hierarchical adhesion structures of geckos. *Mech. Mater.* **2005**, *37*, 275–285. [CrossRef]

coatings

MDPI

Article

Preparation and Photoluminescence of Tungsten Disulfide Monolayer

Yanfei Lv, Feng Huang, Luxi Zhang, Jiaxin Weng, Shichao Zhao * and Zhenguo Ji *

College of Materials & Environmental Engineering, Hangzhou Dianzi University, Hangzhou 310018, China; lvyanfei@hdu.edu.cn (Y.L.); grafengh@163.com (F.H.); 161200006@hdu.edu.cn (L.Z.); weng_jiaxin@126.com (J.W.)

* Correspondence: zhaoshichao@hdu.edu.cn (S.Z.); jizg@hdu.edu.cn (Z.J.); Tel.: +86-571-8771-3538 (S.Z.); +86-571-8771-3535 (Z.J.)

Received: 20 April 2018; Accepted: 24 May 2018; Published: 30 May 2018

Abstract: Tungsten disulfide (WS_2) monolayer is a direct band gap semiconductor. The growth of WS_2 monolayer hinders the progress of its investigation. In this paper, we prepared the WS_2 monolayer through chemical vapor transport deposition. This method makes it easier for the growth of WS_2 monolayer through the heterogeneous nucleation-and-growth process. The crystal defects introduced by the heterogeneous nucleation could promote the photoluminescence (PL) emission. We observed the strong photoluminescence emission in the WS_2 monolayer, as well as thermal quenching, and the PL energy redshift as the temperature increases. We attribute the thermal quenching to the energy or charge transfer of the excitons. The redshift is related to the dipole moment of WS_2.

Keywords: chemical vapor transport deposition; tungsten disulfide; monolayer; photoluminescence

1. Introduction

Tungsten disulfide (WS_2) monolayer, a direct band gap semiconductor, shows strong photoluminescence (PL) emission [1,2]. The PL properties have attracted plenty of research interest.

Yun et al. [3] found the non-uniformity of the PL emission in WS_2 monolayer. The exciton and charge carriers form the exciton complexes. Kim et al. [4] suspected that the PL emission originated from the exciton complexes. The inhomogeneous distribution of the charge carriers resulted in the PL emission non-uniformity. The PL emission energy shifts as the excitation power or sample temperature changes. For example, Gordo et al. found the laser irradiation changes the PL emission by introducing the carrier doping [5]. Rosenberger et al. studied the relationship between the PL intensity and defect density using conductive atomic force microscopy. They found the PL intensity increases with the decrease of the defect density. Furthermore, they proposed that the defects act as the nonradioactive recombination center [6]. The mechanism for PL emission is under exploration. Besides the charge carrier and defects, chemical doping and surface absorption have an effect on the PL emission. Yao et al. and Feng et al. found the PL energy changed as the WS_2 monolayer was immersed into sodium sulfide (Na_2S) solution, or coated with DNA nucleobases [7,8].

Before having full understanding of the PL emission mechanism, successful preparation of WS_2 monolayer is challenging. Under this background, various methods have been applied for the preparation of WS_2 monolayer/layers, such as exfoliation, atomic layer deposition, and pulsed laser deposition [9–11]. Since chemical vapor deposition (CVD) has been successfully applied to the deposition of large-scale two-dimensional materials, such as graphene and hexagonal boron nitride, the CVD method is carried forward for the synthesis of the WS_2 monolayer [12–14]. Among various CVD methods, the sulfurization of tungsten oxide in sulfur vapor is a commonly used CVD method. Gutiérrez et al. deposited a thin film of tungsten oxide on the SiO_2/Si substrate, then heated at 800 °C under argon and sulfur vapor. The lateral size of the WS_2 monolayer with triangular shape is ~15 μm.

The WS_2 showed extraordinary photoluminescence emission at room temperature [2]. Cong et al. reported the growth of WS_2 monolayer on the SiO_2/Si substrate through sulfurization of tungsten oxide powders. The lateral size of triangular WS_2 domain was up to hundreds of micrometers [15]. Before long, Gao et al. [16] reported the growth of WS2 monolayer with millimeter scale using catalyst. They found the photoluminescence (PL) emission intensity at 612.6 nm was 10^3 times stronger than bulk material. Cong et al. [15] and Gao et al. [16] found the PL emission was far stronger in the edge region, which is opposite to that of Gutiérrez's observation. Besides the metal catalyst, WS_2 monolayer growth is also sensitive to the promoter and surface condition. Li et al. found the addition of the alkali metal halides reduces the growth temperature (700–850 °C) [17]. In conclusion, the growth of WS_2 monolayer is sensitive to the growth process. Much research into the preparation and photoluminescence needs to be done.

We recently reported the chemical vapor transport deposition of molybdenum disulfide monolayer [18]. Molybdenum disulfide reacted with the transport agent water vapor to form molybdenum oxide at high temperature. Then, the molybdenum oxide reacted with sulfur and transformed into a molybdenum disulfide monolayer at a low temperature. In this method, water vapor was used as the transport agent and the nucleation promoter. The introduction of water vapor promoted the molybdenum disulfide monolayer growth. WS_2 and molybdenum disulfide are both transition metal dichalcogenides. An interesting question is whether this method can prepare WS_2 monolayer. Here, we show that water vapor can promote the WS_2 monolayer growth. The synthesized WS_2 monolayer shows strong PL emission, and the PL is also sensitive to the temperature. Our findings, therefore, offer important clues to the transition metal dichalcogenides monolayer growth. The PL change couldalso be used as a detector of temperature change.

2. Materials and Methods

WS_2 Layer Synthesis

WS_2 monolayer was prepared by a previously reported chemical vapor transport deposition method, which was used for the growth of molybdenum disulfide monolayer on a silicon substrate with a 300 nm layer of oxide (SiO_2/Si) [18]. WS_2 powders (99.5% purity, Aladdin, Shanghai, China) were used as the precursor. The precursor (0.5 g) was loaded in the center of a one-inch-diameter tube furnace (Hefei Kejing Materials technology Co., Ltd., Hefei, China), and heated to 1000 °C from room temperature in 30 min. The substrate was put downstream of the tube furnace. During the above process, the substrate was heated to 710–850 °C and kept at 710–850 °C for 60 min, before the furnace was turned off and cooled naturally to room temperature. The water vapor was carried into the tube by Ar/H_2 (75 Torr, Ar/H_2 70 sccm) during the whole heating stage.

Optical microscope image was carried out on a Jiangnan MV3000 digital microscope (Nanjing Jiangnan Novel Optics Co., Ltd., Nanjing, China). Tapping mode atomic force microscopy (AFM) was conducted with an Agilent 5500 (Agilent Technologies, Palo Alto, CA, USA) in the air. Raman spectra and photoluminescence (PL) were measured on a home-built micro-Raman setup that consists of a 532 nm solid state laser, a Nikon inverted microscope (Ti eclipse, Nikon, Tokyo, Japan), a long pass edge filter (Semrock, Rochester, NY, USA), and a Raman spectrometer (iHR320, Horiba, Kyoto, Japan) with an attached thermoelectric-cooled CCD camera (Andor Syncerity, Horiba) at room temperature. PL mapping was measured on a Nib400 fluorescence microscope (Nanjing Jiangnan Novel Optics Co., Ltd., Nanjing, China). X-ray diffraction (XRD) was performed on a Thermo ARLXTRA (Thermo Electron, Waltham, MA, USA).

3. Results and Discussion

3.1. WS₂ Monolayer Growth

Figure 1a shows the optical image of the separated triangular WS_2 monolayer grown on the SiO_2/Si substrate in the presence of H_2O vapor. Except for the black circled area which is WS_2 multilayers, other areas with a triangular shape are the WS_2 monolayer. The uniform color contrast indicates that the thickness of the WS_2 monolayer is uniform. The lateral size of the monolayer is up to ca. 38 μm.

Figure 1b shows the PL mapping of the WS_2 monolayer using an excitation wavelength of 485 nm. PL mapping was measured at the same location as the one in Figure 1a (marked with black square). We found the WS_2 monolayer emitted strong PL while the WS_2 multilayer (circled area) gave out no PL emission. Intensive PL emission is due to the direct band gap structure of monolayer. The valence band maximum and the conduction band minimum in the bulk WS_2 occur at different k (wave vector) values in the Brillouin zone. While in the monolayer, the valence band maximum and the conduction band minimum occur at the same position (K point). Therefore, the indirect-to-direct band gap conversion occurs as the thickness of the WS_2 is reduced to a monolayer [19]. The direct band gap is ca. 2.05 eV [20].

Figure 1c shows the PL spectrum of the WS_2 monolayer excited by a 532 nm laser. The peak energy (~2.0 eV) is lower than the band gap, and is considered to be related to the excitons [4]. In this paper, the WS_2 was synthesized by heterogeneous nucleation-and-growth process (discussed in the section of the mechanism below). The heterogeneous nucleation could promote the crystal growth as well as the introduction of crystal defects. The crystal defects play a significant role in the PL emission. Lattice defects act as nonradiative recombination centers in general semiconductor.

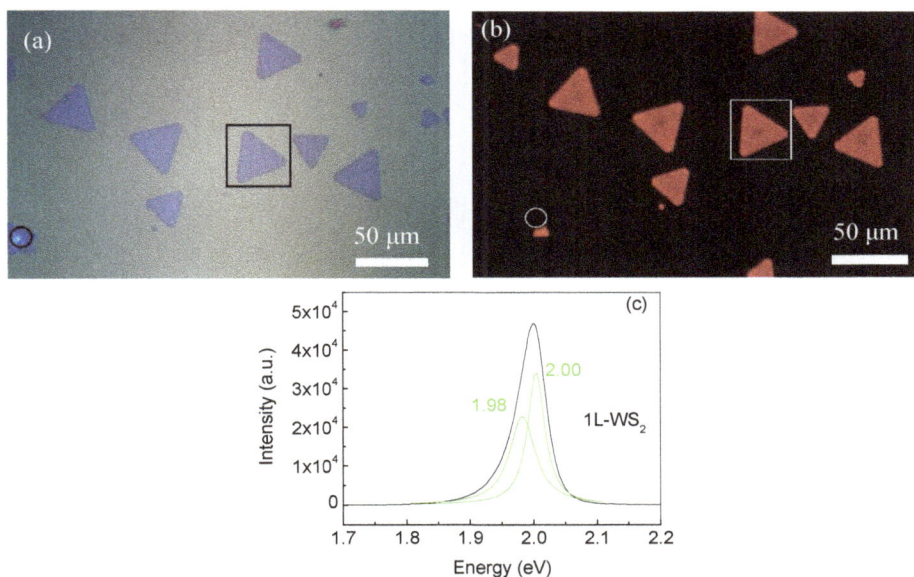

Figure 1. (a) Optical microscopy image of tungsten disulfide monolayer. (b) Corresponding photoluminescence (PL) mapping. The multilayer was circled in (a,b). (c) Photoluminescence (PL) spectrum of the tungsten disulfide monolayer, which was taken from the monolayer region labelled with a box in (a)/(b). The spectrum was fitted with Lorentzian function (green lines). The peaks at 2.00 and 1.98 eV are due to the transition of charged and defect-related excitons, respectively. The scale bars represent 50 μm.

However, recent investigations revealed that the lattice defects can promote the PL emission by forming charged exciton and defect-bound excitons with neutral excitons [4,21]. We fit the strong PL peak at ca. 2.0 eV with Lorentzian lineshape functions, as shown in Figure 1c. The two components at 2.00 eV and 1.98 eV are possibly due to the transition of charged excitons and defect-bound excitons, respectively. The intensive PL and single emission peak are characteristics for WS_2 monolayer, and once again, testify that the thickness of the WS_2 is monolayer [2,16].

Figure 2a shows the AFM image of the WS_2 flake edge within the square in Figure 1a. The height of the edge is 1.0 ± 0.1 nm, which is larger than the theoretical value (0.6 nm) of the monolayer thickness [22]. Considering the tip–substrate interaction and the surface adsorbate, the flake should be a monolayer [18].

Figure 3 shows the Raman spectrum of the WS_2 monolayer. The anti-symmetric peak E^1_{2g} at ca. 357 cm^{-1} can be fitted with Lorentzian function whose three subpeaks are at 343, 351, and 356 cm^{-1}, corresponding to the in-plane vibrational $E^1_{2g}(M)$ mode, the longitudinal acoustic phonon 2LA(M) mode, and the in-plane vibrational $E^1_{2g}(\Gamma)$, respectively. The peak at 418 cm^{-1} (A_{1g}) is assigned to the out-of-plane vibrational mode of two sulfur atoms [4,20].

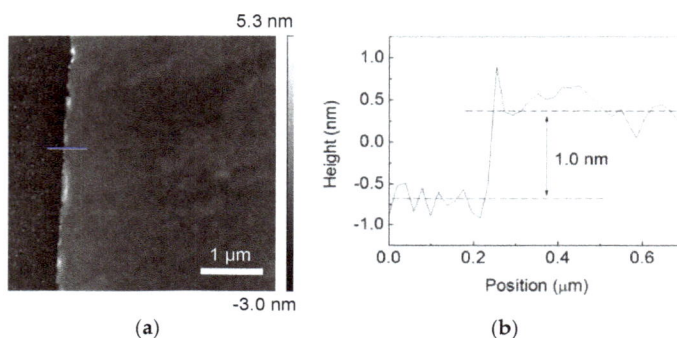

Figure 2. Atomic force microscopy (AFM) image (**a**) and the corresponding cross-section (**b**) (along the blue line marked in (**a**)) of monolayer WS_2 grown on the SiO_2/Si substrate.

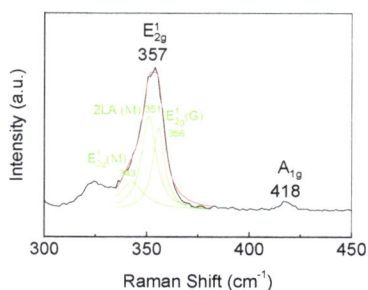

Figure 3. Typical Raman spectrum of the WS_2 monolayer.

To study the growth mechanism of the WS_2 monolayer, we analyzed the residue of the precursor by X-ray diffraction (XRD). Figure 4 shows the X-ray diffraction (XRD) of the precursor after heating in water vapor for more than 20 h at 1000 °C. Peaks of tungsten oxides are found in the XRD spectrum. Therefore, the WS_2 reacted with water vapor to form tungsten oxides, which further sulfurized and transformed into the WS_2 monolayer. Tungsten oxides acted as the heterogeneous nucleation during the growth process.

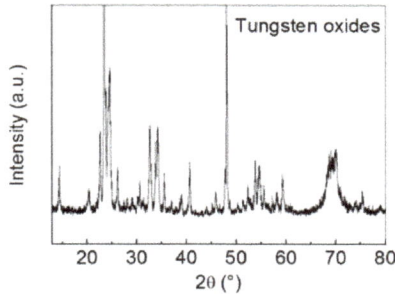

Figure 4. X-ray diffraction (XRD) of WS_2 powder after annealing in Ar/H_2 and H_2O vapor at 1000 °C for 20 h. The peak positions are indexed to tungsten oxides $W_{18}O_{49}$ (JCPDS No. 05-0392 and No. 65-5468).

3.2. Temperature Dependence of the PL

For the investigation of the in situ PL under different temperatures, we heated the WS_2 monolayer from 15 to 63 °C by a temperature-controlled heater.

Figure 5a displays the PL spectra as a function of the temperature. The PL intensity decreases with the increasing temperature as shown in Figure 5b. The intensity measured at 63 °C was reduced to 6% of the initial value. The curve in Figure 5b is fitted by an exponential function (Equation (1)).

$$I\,(T) = 3.5 \times 10^7 \exp(-0.085T) + 5.9 \times 10^4 \qquad (1)$$

where, the I is the PL intensity and T is the temperature. The reason behind the exponential behavior will be reported in a future work. Note that the value measured at 18 °C is larger than that at 15 °C. We suspect the exception is possibly due to the experimental error.

Kim et al. [4] found the PL intensity of the WS_2 monolayer was reduced after annealing at 800 °C. They suggested the decrease of the PL intensity is attributed to the reduction of the number of the excitons. Su et al. observed the thermal quenching of the WS_2 monolayer in the range of room temperature to 400 °C [23]. They attributed the thermal quenching to the thermal activation of nonradiative recombination processes. In addition, they found the strong interaction between the WS_2 and the substrate could quench the PL, and pointed out that the interfacial effect may play an important role in the understanding of the 2D materials behavior. Besides the abovementioned possible mechanisms, we suspect another possible reason for the thermal quenching is related to the dipole moment.

In addition to the quenching, the red shift of the PL energy is clearly observed. The PL peak energy reduced from 1.98 to 1.92 eV as temperature increased from 15 to 63 °C with an average rate of ~-1.2 meV/°C. Strain is reported as a possible reason for the energy shift [23]. If there is strain in the sample, the Raman peak position should shift. Figure 6a shows the Raman spectra of WS_2 monolayer under different temperatures. The Raman spectra were normalized using the Si peak. However, we found no obvious Raman shift of the A_{1g} mode in Figure 6a. The Raman spectra were measured near room temperature, and the small variation of the temperature may not cause an evident strain, even with a Raman shift. Therefore, we are not sure whether the strain has an effect on the PL energy shift. However, the Raman data in Figure 6b clearly show that the intensity of A_{1g} has a positive proportional linear function with the temperature, indicating the out-of-plane dipole moment increases with the increase of the temperature. As the dipole moment has an effect on the exciton energy, we suspected the increased dipole moment resulted in the PL energy redshift [24]. Furthermore, the increased dipole moment may promote the energy transfer or charge transfer from the excitons to substrate or defects in WS_2, resulting the PL quenching.

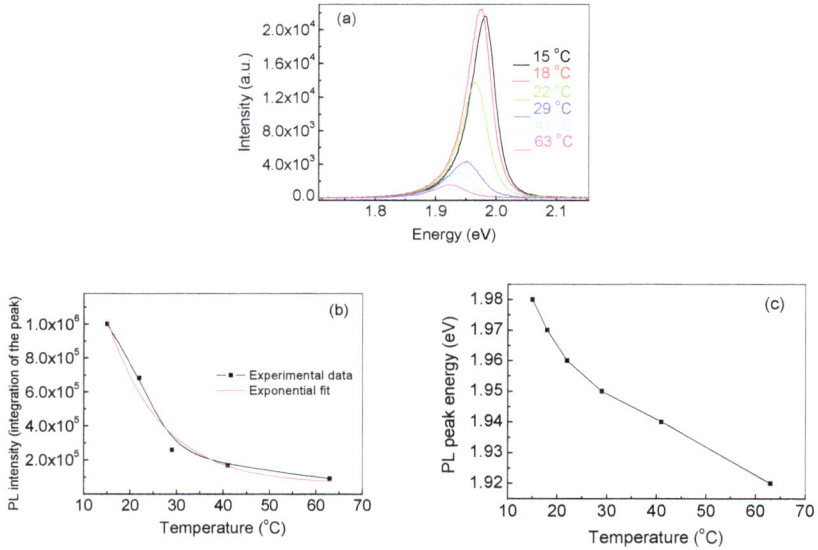

Figure 5. (**a**) Photoluminescence (PL) spectra of the WS$_2$ monolayer measured at different temperatures, showing the decrease of intensity and redshift of the PL energy with the rising temperatures. (**b**) Integrated PL intensity as a function of temperature. The PL intensity was calculated by integration from 1.8 to 2.1 eV. The curve was fit with an exponential function (red line). (**c**) Temperature dependence of the PL peak energy.

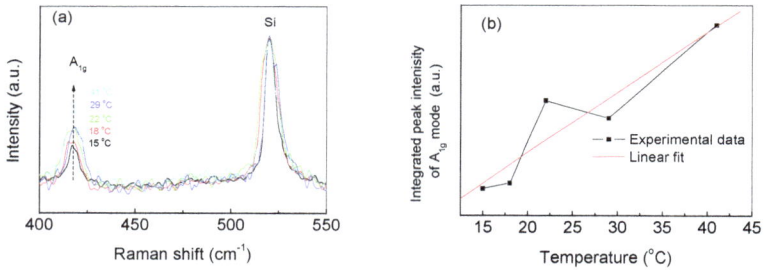

Figure 6. (**a**) Raman spectra of the WS$_2$ monolayer at different temperatures. The Raman spectra were normalized using the Si peak. (**b**) Integrated Raman intensity of A_{1g} mode as a function of temperature. The intensity was calculated by integration from 400 to 434 cm^{-1}. The curve was fit with a linear function (red line).

4. Conclusions

In summary, we have successfully prepared WS$_2$ monolayer in the presence of H$_2$O vapor deposition, suggesting that H$_2$O vapor could be used as a transport agent for the growth of WS$_2$. The AFM, Raman, and PL revealed that monolayer WS$_2$ with triangular shape was formed. Using XRD, we showed that tungsten oxides were formed by the reaction between WS$_2$ and water, which promoted the growth of WS$_2$. By heating the WS$_2$ monolayer, we showed the PL energy and intensity are sensitive to the temperature. The increased dipole moment and energy or charge transfer may be possible reasons for the PL spectra change.

Coatings **2018**, *8*, 205

Author Contributions: Y.L. conceived and designed the experiments and analyzed the data; F.H., L.Z. and J.W. performed the experiments; Z.J. analyzed the data; S.Z. wrote the paper.

Funding: This work was funded by the Natural Science Foundation of Zhejiang Province, China Projects (LY16E020008) and Chinese NSF Projects (61106100).

Conflicts of Interest: The authors declare no conflict of interest.

References

1. Zhang, Y.; Zhang, Y.F.; Ji, Q.Q.; Ju, J.; Yuan, H.T.; Shi, J.P.; Gao, T.; Ma, D.L.; Liu, M.X.; Chen, Y.B.; et al. Controlled growth of high-quality monolayer WS$_2$ layers on sapphire and imaging its grain boundary. *ACS Nano* **2013**, *7*, 8963–8971. [CrossRef] [PubMed]
2. Gutierrez, H.R.; Perea-Lopez, N.; Elias, A.L.; Berkdemir, A.; Wang, B.; Lv, R.; Lopez-Urias, F.; Crespi, V.H.; Terrones, H.; Terrones, M. Extraordinary room-temperature photoluminescence in triangular WS$_2$ monolayers. *Nano Lett.* **2013**, *13*, 3447–3454. [CrossRef] [PubMed]
3. Yun, S.J.; Chae, S.H.; Kim, H.; Park, J.C.; Park, J.H.; Han, G.H.; Lee, J.S.; Kim, S.M.; Oh, H.M.; Seok, J.; et al. Synthesis of centimeter-scale monolayer tungsten disulfide film on gold foils. *ACS Nano* **2015**, *9*, 5510–5519. [CrossRef] [PubMed]
4. Kim, M.S.; Yun, S.J.; Lee, Y.; Seo, C.; Han, G.H.; Kim, K.K.; Lee, Y.H.; Kim, J. Biexciton emission from edges and grain boundaries of triangular WS$_2$ monolayers. *ACS Nano* **2016**, *10*, 2399–2405. [CrossRef] [PubMed]
5. Gordo, V.O.; Balanta, M.A.G.; Gobato, Y.G.; Covre, F.S.; Galeti, H.V.A.; Iikawa, F.; Couto, O.D.D., Jr.; Qu, F.; Henini, M.; Hewak, D.W.; et al. Revealing the nature of low-temperature photoluminescence peaks by laser treatment in Van der Waals epitaxially grown WS$_2$ monolayers. *Nanoscale* **2018**, *10*, 4807–4815. [CrossRef] [PubMed]
6. Rosenberger, M.R.; Chuang, H.J.; McCreary, K.M.; Li, C.H.; Jonker, B.T. Electrical characterization of discrete defects and impact of defect density on photoluminescence in monolayer WS$_2$. *ACS Nano* **2018**, *12*, 1793–1800. [CrossRef] [PubMed]
7. Yao, H.; Liu, L.; Wang, Z.; Li, H.; Chen, L.; Pam, M.E.; Chen, W.; Yang, H.Y.; Zhang, W.; Shi, Y. Significant photoluminescence enhancement in WS$_2$ monolayers through Na$_2$S treatment. *Nanoscale* **2018**, *10*, 6105–6112. [CrossRef] [PubMed]
8. Feng, S.; Cong, C.X.; Peimyoo, N.; Chen, Y.; Shang, J.Z.; Zou, C.J.; Cao, B.C.; Wu, L.S.; Zhang, J.; Eginligil, M.; et al. Tunable excitonic emission of monolayer WS$_2$ for the optical detection of DNA nucleobases. *Nano Res.* **2018**, *11*, 1744–1754. [CrossRef]
9. Xu, D.Y.; Xu, P.T.; Zhu, Y.Z.; Peng, W.C.; Li, Y.; Zhang, G.L.; Zhang, F.B.; Mallouk, T.E.; Fan, X.B. High yield exfoliation of WS$_2$ crystals into 1–2 layer semiconducting nanosheets and efficient photocatalytic hydrogen evolution from WS$_2$/CdS nanorod composites. *ACS Appl. Mater. Interfaces* **2018**, *10*, 2810–2818. [CrossRef] [PubMed]
10. Groven, B.; Heyne, M.; Mehta, A.N.; Bender, H.; Nuytten, T.; Meersschaut, J.; Conard, T.; Verdonck, P.; Van Elshocht, S.; Vandervorst, W.; et al. Plasma-enhanced atomic layer deposition of two-dimensional WS$_2$ from WF$_6$, H$_2$ plasma, and H$_2$S. *Chem. Mater.* **2017**, *29*, 2927–2938. [CrossRef]
11. Sahu, R.; Radhakrishnan, D.; Vishal, B.; Negi, D.S.; Sil, A.; Narayana, C.; Datta, R. Substrate induced tuning of compressive strain and phonon modes in large area MoS$_2$ and WS$_2$ van der Waals epitaxial thin films. *J. Cryst. Growth* **2017**, *470*, 51–57. [CrossRef]
12. Bae, S.; Kim, H.; Lee, Y.; Xu, X.F.; Park, J.S.; Zheng, Y.; Balakrishnan, J.; Lei, T.; Kim, H.R.; Song, Y.I.; et al. Roll-to-roll production of 30-inch graphene films for transparent electrodes. *Nat. Nanotechnol.* **2010**, *5*, 574–578. [CrossRef] [PubMed]
13. Zhao, S.C.; Surwade, S.P.; Li, Z.T.; Liu, H.T. Photochemical oxidation of CVD-grown single layer graphene. *Nanotechnology* **2012**, *23*, 355703. [CrossRef] [PubMed]
14. Zhao, S.C.; Zhou, F.; Li, Z.T.; Liu, H.T. Effect of precursor purity and flow rate on the CVD growth of hexagonal boron nitride. *J. Alloys Compd.* **2016**, *688*, 1006–1012. [CrossRef]
15. Cong, C.X.; Shang, J.Z.; Wu, X.; Cao, B.C.; Peimyoo, N.; Qiu, C.; Sun, L.T.; Yu, T. Synthesis and optical properties of large-area single-crystalline 2D semiconductor WS$_2$ monolayer from chemical vapor deposition. *Adv. Opt. Mater.* **2014**, *2*, 131–136. [CrossRef]

16. Gao, Y.; Liu, Z.B.; Sun, D.M.; Huang, L.; Ma, L.P.; Yin, L.C.; Ma, T.; Zhang, Z.Y.; Ma, X.L.; Peng, L.M.; et al. Large-area synthesis of high-quality and uniform monolayer WS_2 on reusable Au foils. *Nat. Commun.* **2015**, *6*, 8569. [CrossRef] [PubMed]

17. Li, S.; Wang, S.; Tang, D.; Zhao, W.; Xu, H.; Chu, L.; Bando, Y.; Golberg, D.; Eda, G. Halide-assisted atmospheric pressure growth of large WSe_2 and WS_2 monolayer crystals. *Appl. Mater. Today* **2015**, *1*, 60–66. [CrossRef]

18. Zhao, S.C.; Weng, J.X.; Jin, S.Z.; Lv, Y.F.; Ji, Z.G. Chemical vapor transport deposition of molybdenum disulfide layers using H_2O vapor as the transport agent. *Coatings* **2018**, *8*, 78. [CrossRef]

19. Ma, Y.D.; Dai, Y.; Guo, M.; Niu, C.W.; Lu, J.B.; Huang, B.B. Electronic and magnetic properties of perfect, vacancy-doped, and nonmetal adsorbed $MoSe_2$, $MoTe_2$ and WS_2 monolayers. *Phys. Chem. Chem. Phys.* **2011**, *13*, 15546–15553. [CrossRef] [PubMed]

20. Peimyoo, N.; Shang, J.Z.; Cong, C.X.; Shen, X.N.; Wu, X.Y.; Yeow, E.K.L.; Yu, T. Nonblinking, intense two-dimensional light emitter: Mono layer WS_2 triangles. *ACS Nano* **2013**, *7*, 10985–10994. [CrossRef] [PubMed]

21. Chow, P.K.; Jacobs-Gedrim, R.B.; Gao, J.; Lu, T.M.; Yu, B.; Terrones, H.; Koratkar, N. Defect-induced photoluminescence in monolayer semiconducting transition metal dichalcogenides. *ACS Nano* **2015**, *9*, 1520–1527. [CrossRef] [PubMed]

22. Stier, A.V.; McCreary, K.M.; Jonker, B.T.; Kono, J.; Crooker, S.A. Exciton diamagnetic shifts and valley Zeeman effects in monolayer WS_2 and MoS_2 to 65 Tesla. *Nat. Commun.* **2016**, *7*, 10643. [CrossRef] [PubMed]

23. Su, L.Q.; Yu, Y.F.; Cao, L.Y.; Zhang, Y. Effects of substrate type and material-substrate bonding on high-temperature behavior of monolayer WS_2. *Nano Res.* **2015**, *8*, 2686–2697. [CrossRef]

24. Motta, C.; Mandal, P.; Sanvito, S. Effects of molecular dipole orientation on the exciton binding energy of $CH_3NH_3PbI_3$. *Phys. Rev. B* **2016**, *94*, 045202. [CrossRef]

Article

Chemical Vapor Transport Deposition of Molybdenum Disulfide Layers Using H_2O Vapor as the Transport Agent

Shichao Zhao *, Jiaxin Weng, Shengzhong Jin, Yanfei Lv and Zhenguo Ji *

College of Materials & Environmental Engineering, Hangzhou Dianzi University, Hangzhou 310018, China; weng_jiaxin@126.com (J.W.); wk_jsz@163.com (S.J.); lvyanfei@hdu.edu.cn (Y.L.)
* Correspondence: zhaoshichao@hdu.edu.cn (S.Z.); jizg@hdu.edu.cn (Z.J.)

Received: 8 January 2018; Accepted: 14 February 2018; Published: 21 February 2018

Abstract: Molybdenum disulfide (MoS_2) layers show excellent optical and electrical properties and have many potential applications. However, the growth of high-quality MoS_2 layers is a major bottleneck in the development of MoS_2-based devices. In this paper, we report a chemical vapor transport deposition method to investigate the growth behavior of monolayer/multi-layer MoS_2 using water (H_2O) as the transport agent. It was shown that the introduction of H_2O vapor promoted the growth of MoS_2 by increasing the nucleation density and continuous monolayer growth. Moreover, the growth mechanism is discussed.

Keywords: chemical vapor transport deposition; molybdenum disulfide; monolayer; water; mechanism

1. Introduction

Molybdenum disulfide (MoS_2) layers, having unique optical and electrical properties, have attracted extensive interest in the fields of energy generation, electronics, and sensors [1–7]. The growth of large-scale, high-quality MoS_2 layers targeted for silicon integrated device fabrication is still challenging. Vapor deposition has been the predominant method for the growth of large-scale, continuous MoS_2 monolayer or few layers films in recent years [8–10]. Molybdenum oxides and sulfur are generally used as precursors of MoS_2. For example, Lee et al. heated MoO_3 powder in sulfur vapor and obtained MoS_2 monolayer and multi-layer films [11]. In this method, MoO_3 was first reduced by sulfur vapor to form MoO_{3-x}, which was then further reacted with sulfur vapor to form MoS_2 [12]. MoO_3 acted as a nucleation center promoting crystal growth as well as the introduction of crystal defects. The introduction of defects plays two important roles; one is to promote nucleation for multi-layer growth, and the other is to tailor the electrical properties [13–16]. The MoS_2 domains grown with this method showed different morphologies, e.g. triangle, hexagon, three-point star, as a function of the different atomic ratio of sulfur to molybdenum [17,18].

MoS_2 powder is another commonly used starting material. Wu et al. [19] heated MoS_2 powder at 900 °C in the center of a tube furnace and obtained a MoS_2 monolayer on an insulating substrate downstream of the precursor in a lower temperature zone (~650 °C). The usage of single-precursor MoS_2 powder as the source of Mo and S avoided the introduction of impurities and heterogeneous nucleation during the growth of MoS_2 flakes. Therefore, the MoS_2 monolayer showed a regular triangular shape and high optical quality. This vapor-solid growth method is suitable for the deposition of high-quality monolayer single crystal flake. However, in our recent work, we found that the nucleation is difficult to initiate and the growth temperature window is very narrow, ca. ~50 °C [20]. These issues could be attributed to the low vapor pressure of MoS_2 powder. The chemical vapor transport method was generally used for the growth of crystals with a solid precursor that has low vapor pressure. For example, Pisoni et al. reported the growth of MoS_2 single crystals using I_2, Br_2, and $TeCl_4$ as transport agents [21]. The transport agent converts MoS_2 into high vapor pressure intermediates, which undergo the reverse reaction to deposit MoS_2 onto the substrate. However, the vapor transport agents used in this study are highly toxic and reactive, which could limit their widespread use.

To overcome this limitation, we have investigated ways to improve the nucleation density of MoS_2 using various additives. In this paper, we report the chemical vapor transport growth behavior of MoS_2 monolayer or multi-layer films by using MoS_2 powder as the precursor and water (H_2O) vapor as the transport agent. In the nucleation stage, H_2O vapor was introduced into the deposition system and acted as a chemical transport agent. Our mechanistic study suggests that water reacted with MoS_2 to form MoO_2, which promoted the nucleation of MoS_2. In the previously mentioned growth methods, the sulfur comes from the sublimation of sulfur or MoS_2 powder and the sulfur flow rate is out of control. Here, the sulfur was formed through the reaction of MoS_2 and water, which provides us a possible way to adjust the sulfur flow rate by controlling the water vapor flow rate. In the second stage, H_2O vapor was cut off and MoS_2 continuously grew through a simple physical vapor transport process. This novel approach combined the heterogeneous nucleation and homogeneous growth to control the crystal size and thickness of the MoS_2 layer. The thickness of the MoS_2 film obtained ranged from a monolayer to multiple layers. The lateral size of the single-crystal domain is up to 300 μm.

2. Materials and Methods

MoS₂ Layers Synthesis

MoS_2 was prepared by modifying a previously reported vapor deposition method using a silicon wafer with a 300-nm layer of oxide (SiO_2/Si) as the substrate [19]. The schematic of the vapor deposition setup is shown in Figure 1. MoS_2 powder (99.5% purity, Aladdin, Shanghai, China) was used as the precursor. Before use, the precursor (0.5 g) was loaded into a small quartz glass boat (70 mm in length) and put in the center of the tube furnace (1 inch in diameter, Hefei Kejing Materials technology Co. Ltd., Hefei, China). Before growth, the precursor was flushed under Ar/H_2 (70 sccm, H_2 5%, total pressure of 75 Torr. sccm: standard cubic centimeter per minute) for 10 min at room temperature to remove the air and water absorbed on the precursor. The substrate was put downstream close to the furnace wall.

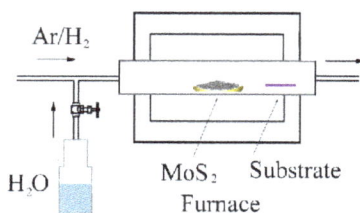

Figure 1. Schematic illustration of the MoS_2 growth setup.

For the MoS_2 growth, the precursor was heated to 1000 °C from room temperature in 30 min under Ar/H_2(75 Torr, Ar/H_2 70 sccm) and kept at 1000 °C for 1 h. The furnace was then turned off and cooled from 1000 °C to room temperature. During the above heating process, the temperature of the substrate ranged from 710 to 850 °C. The water (H_2O) vapor was introduced into the furnace by turning on/off the water valve, which connects the water tube and the Ar/H_2 inlet. For typical growth, the water valve was kept open during the whole heating stage. For studies on the influence of H_2O on the growth of MoS_2, we kept the valve open during the heating stage but limited the water exposure during the synthesis.

Optical microscope imaging of the sample was conducted with a Jiangnan MV3000 digital microscope (Nanjing Jiangnan Novel Optics Co. Ltd., Nanjing, China). Tapping mode atomic force microscopy (AFM) was performed on an Agilent 5500 (Agilent Technologies, Palo Alto, CA, USA) in air. Raman spectrum and photoluminescence (PL) were acquired on a Renishawin Via micro-Raman spectroscope (Renishaw, London, UK) with a 532 nm solid-state laser at room temperature. X-ray diffraction (XRD) was carried on a Thermo ARLXTRA (Thermo Electron, Waltham, USA) and ultraviolet visible diffuse reflection spectroscopy (UV-Vis DRS, not including specular reflection) was performed on Shimadzu MPC-3100 (Shimadzu, Tokyo, Japan) with an integrating sphere.

3. Results and Discussion

3.1. MoS$_2$ Flakes Grown in the Presence of H$_2$O Vapor

MoS$_2$ flakes were prepared on the substrate using H$_2$O and MoS$_2$ powder as illustrated in Figure 2a. Figure 2b shows the separated triangular MoS$_2$ flakes grown on the SiO$_2$/Si substrate with the H$_2$O vapor valve kept open during the heating of the furnace and growth of the MoS$_2$ flakes. The thickness of the flakes ranged from monolayer to multiple layers. The triangles in dim and uniform color indicate the uniform monolayer MoS$_2$. The bright color triangles are attributed to multi-layer MoS$_2$ with a pyramid-shape structure. The flake lateral size ranged from ca. 20 to 40 μm.

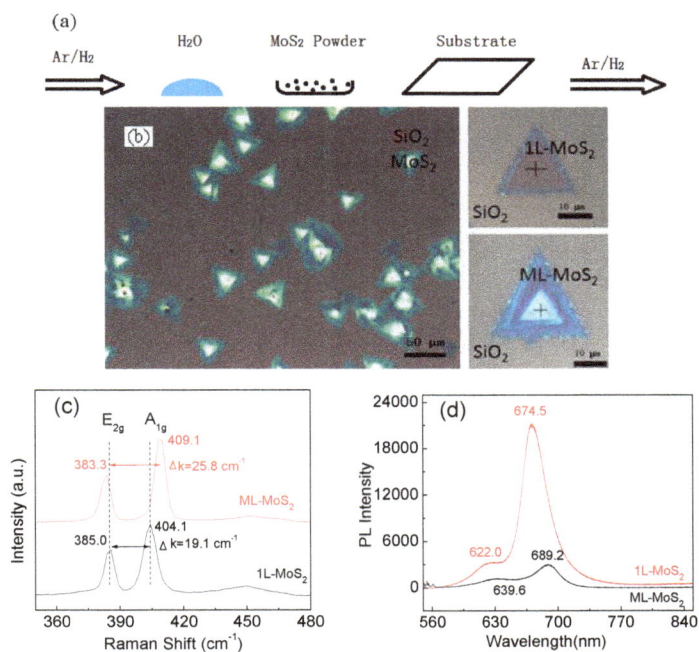

Figure 2. (**a**) Schematic illustration of the MoS$_2$ growth using H$_2$O and MoS$_2$ powder; (**b**) Optical images of MoS$_2$ grown on a SiO$_2$/Si substrate with H$_2$O vapor for 10 min; (**c**) Typical Raman spectra and (**d**) Photoluminescence (PL) spectra of the monolayer (1L-MoS$_2$) and multi-layer MoS$_2$ (ML-MoS$_2$) flakes according images shown in Figure 2b.

The success of the growth and thickness of the MoS$_2$ flakes were confirmed by Raman spectroscopy. Figure 2c displays the typical Raman spectra of monolayer and multi-layer MoS$_2$ flakes corresponded to the images in Figure 2b. The E_{2g} and A_{1g} modes of MoS$_2$ were observed. The frequency difference between the E_{2g} and A_{1g} mode is thickness-dependent. With the increase of the layer number, the frequency difference valve will increase. The E_{2g} and A_{1g} peaks positions are at 385.0 cm^{-1} and 404.1 cm^{-1} (383.3 cm^{-1} and 409.1 cm^{-1}) with a frequency difference of 19.9 cm^{-1} (25.8 cm^{-1}), indicating that the thickness of flakes is monolayer (multi-layer) [22].

Besides Raman spectra, PL is generally used for the identification of the thickness of the MoS_2. Mak et al. studied the relationship between the PL quantum yield and layer number. They found that the PL quantum yield drops quickly with the increase of the layer number. Bulk MoS_2 is an indirect-gap semiconductor showing negligible PL. Few-layer MoS_2 shows weak PL due to the confinement effects. Monolayer MoS_2 is a direct-gap semiconductor giving out bright PL [23]. Figure 2d shows the typical photoluminescence spectra (PL) both of the monolayer and multi-layer MoS_2 flakes corresponded to the images in Figure 2b. The excitation wavelength was 532 nm. The PL peaks of monolayer MoS_2 are located at 674.5 nm and 622 nm, which are attributed to the A1 and B1 direct excitonic transition emission of the MoS_2 monolayer, respectively [9,17,24]. We observed that the PL intensity of the monolayer is much stronger than that of the multi-layer sample.

3.2. Effect of H_2O Vapor on the MoS_2 Layers Growth

To investigate the effect of H_2O vapor on the MoS_2 growth, we limited the time the synthesis was exposed to water vapor. After the precursors were heated to 1000 °C, the water valve was closed after a fixed amount of time during the growth stage: Figure 3a–d 0 min (least water exposure), Figure 3e–h 10 min, Figure 3i–l 20 min, and Figure 3m–p 60 min (most water exposure).

Shown in Figure 3a–d, the shape of the MoS_2 prepared without the introduction of H_2O is a separated island. Meanwhile in Figure 3e,i–k,m–o, continuous, large-area MoS_2 films were observed. This may be due to the presence of H_2O vapor, which enhanced the diffusibility of molybdenum and sulfur atoms at domain boundaries, resulting in the continuous growth of monolayer MoS_2. The large optical contrast in Figure 3e–p indicates the formation of multiple layers and/or clusters, which may be due to the formation of high heterogeneous nucleation density and the Stranski-Krastanov growth mode. The formation of heterogeneous nucleation will be discussed below. Besides the continuous film obtained as described above, the domain size of MoS_2 prepared with H_2O (shown in Figure 3f–h,p) was larger than those prepared without H_2O (shown in Figure 3b,c). Figure 4 shows the magnified optical image of the same sample that is shown in Figure 3e. The lateral size of the triangle-shaped MoS_2 flakes ranges from 24 μm to 372 μm. The average lateral size of the MoS_2 flakes prepared without H_2O introduction was 13 ± 6 μm, while the average size increased to 159 ± 80 μm based on the statistical calculation of the size of the isolated flakes shown in Figure 3a–h, respectively.

The water introduction also has an effect on the thickness of MoS_2 flakes. Based on the frequency difference (24.7 cm^{-1}, Figure S1) between the E_{2g} and A_{1g} mode of MoS_2 and uniform color contrast, we can conclude that the MoS_2 flakes prepared without water exposure in Figure 3a–d is multi-layer. In contrast, in those samples prepared in the presence of water (Figure 3e–p), monolayer MoS_2 was observed (as discussed at the end of Section 3.2). Therefore, the introduction of water can reduce the thickness of the MoS_2 flakes.

It is reported that water molecules and carbon atoms can intercalate between the two-dimensional material and the substrate [25–27]. Although we do not have enough evidence to show the presence of the water intercalation in our sample at high growth temperature (710 °C to 850 °C), we suspect that the molecular structure of water vapor possibly intercalates into the interlayer of the MoS_2 flakes or the interface between the MoS_2 and SiO_2/Si substrate, which affects the absorption, desorption, and diffusion of the precursor atoms and even the final monolayer growth.

From Figure 3m–o, we can observe some bright features. The white spots are multi-layer MoS_2. The area with green and yellow color we suspected to be amorphous MoS_2, MoO_2, or even organic contamination. To reduce the organic contamination, the silicon wafer substrate was cleaned with hot piranha solution (7:3 concentrated H_2SO_4:35% H_2O_2) for 10 min at room temperature, and the vapor deposition system was flushed under Ar/H_2 to remove air-borne contamination before MoS_2 growth.

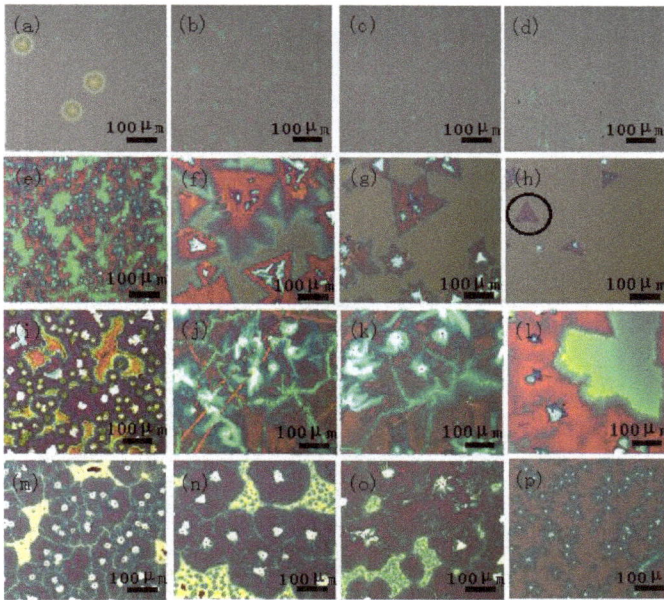

Figure 3. Optical images of the MoS$_2$ grown on a SiO$_2$/Si substrate with varying the amount of H$_2$O vapor released into the furnace. The images in each row are of the same sample but measured at different areas. From the left to right, the deposition temperature decreases as a result of the differences in the location. The amount of H$_2$O vapor into the furnace is controlled by adjusting the length of time that the H$_2$O valve is open: (**a–d**) 0 min; (**e–h**) 10 min; (**i–l**) 20 min; and (**m–p**) 60 min. For (**f,j,k**), we intentionally scratched the sample to show the contrast between the MoS$_2$ film and the SiO$_2$/Si substrate (bright orange color). The scale bars represent 100 µm.

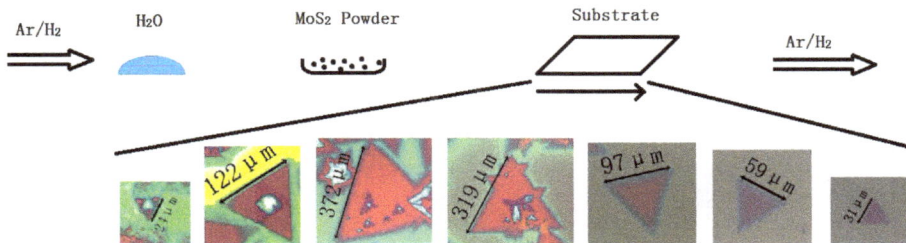

Figure 4. Magnified optical images of the same sample that is shown in Figure 3e. The images were measured at different locations. From left to right, the growth temperature was gradually decreasing.

Figure 5 shows the typical UV-Vis DRS of the MoS$_2$ film corresponding to the images shown in Figure 3e–h. The UV-Vis DRS peaks at 665 and 610 nm match the two PL emission peaks (Figure 2d), and are due to the characteristic A1 and B1 direct excitonic transitions of MoS$_2$, respectively [28].

AFM is a commonly used technique for two-dimensional material thickness measurement. Here, we conducted multiple scans of the thickness of the monolayer MoS$_2$ at an edge of the MoS$_2$ flake by AFM. Figure 6 shows the typical AFM image of the edge of the monolayer MoS$_2$ triangle shown circled in black in Figure 3h. A straight trench with a width of ca. 150 nm was observed on the substrate surface, which divided the substrate into two sections. The bottom of the trench is the SiO_2/Si substrate. The left side of the trench is MoS_2 particles, and the right side of the trench is monolayer MoS_2. The thickness of the MoS_2 flake is 0.9 ± 0.1nm (Figure 6, Figure S2, and Table S1), indicating that the flake is monolayer. This thickness value, although it significantly deviates from the expected thickness of monolayer MoS_2 (0.615 nm), is consistent with other AFM measurements of single-layer MoS_2 deposited on a SiO_2 substrate [29,30]. In fact, the discrepancy that the measured value by AFM is larger than the theoretical value is common phenomena in the measurement of the thickness of two-dimensional monolayer materials, such as graphene [31]. The discrepancy was attributed to the instrument offset due to tip-substrate interaction as well as adsorbed molecules between the monolayer and the SiO_2 substrate [26,31]. From the AFM image in Figure 6, we can see that monolayer MoS_2 is smooth and continuous. We measured the root-mean-square (RMS) surface roughness over a 1 μm × 1 μm area. The RMS was 0.22 nm. The trench is probably formed through the rapid diffusion of MoS_2 nucleation along the direction perpendicular to the domain edge. Detailed study of the trench will be reported in future study. In addition to the trench, there are also many white particles on the surface of the MoS_2 flake. These particles should be MoS_2 formed during the growth of the MoS_2 flake, or even contaminations formed during the transport of the sample.

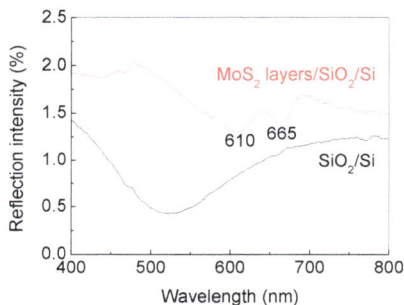

Figure 5. Typical ultraviolet visible diffuse reflection spectroscopy (UV-Vis DRS) of MoS_2 corresponding to the images in Figure 3b.

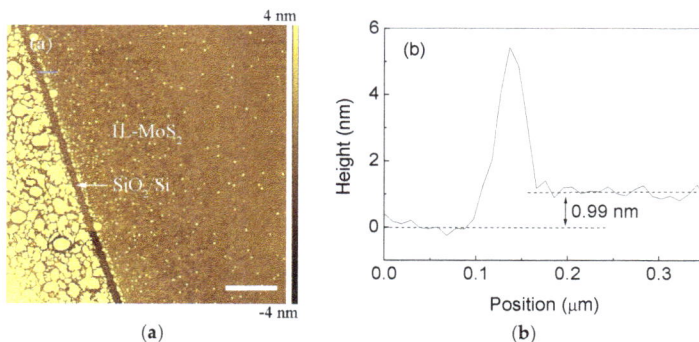

Figure 6. Atomic force microscopy (AFM) image (**a**) and cross-section (**b**) (along the blue line marked in (**a**)) of monolayer MoS_2 grown on the SiO_2/Si substrate corresponding to the images in Figure 3h. The scale bars represent 1 μm in AFM image.

3.3. Mechanism of MoS$_2$ Growth in the Presence of H$_2$O Vapor

The results obtained in Figure 3 suggest that H$_2$O vapor promoted the growth of MoS$_2$ film. We hypothesized that the H$_2$O vapor reacted with MoS$_2$ powder to give molybdenum oxide. Then the molybdenum oxide evaporated and deposited on the substrate, acting as heterogeneous nucleation center, from which the molybdenum oxide reacted with sulfur at a lower temperature and transformed into the MoS$_2$ layer [12]. The following reactions should have occurred during the growth of MoS$_2$ [32]:

$$MoS_2 + 4H_2O \overset{1000\,°C}{\rightarrow} MoO_2 + H_2S + SO_2 + 3H_2 \tag{1}$$

$$2H_2S + SO_2 \overset{\Delta}{\rightarrow} 3S + 2H_2O \tag{2}$$

$$MoO_2 + 2S + 2H_2 \overset{\Delta}{\rightarrow} MoS_2 + 2H_2O \tag{3}$$

To verify this hypothesis experimentally, we used XRD to test the composition of the precursor annealed at 1000 °C for 20 h in H$_2$O vapor and H$_2$/Ar atmosphere. We indeed found that all of the XRD peaks in Figure 7 were indexed according to the monoclinic molybdenum dioxide (MoO$_2$) (JCPDS NO. 00-032-0671). This result agrees with our hypothesis that molybdenum oxide was formed. The growth process essentially is a chemical vapor transport process. The H$_2$O vapor acts as transport agent.

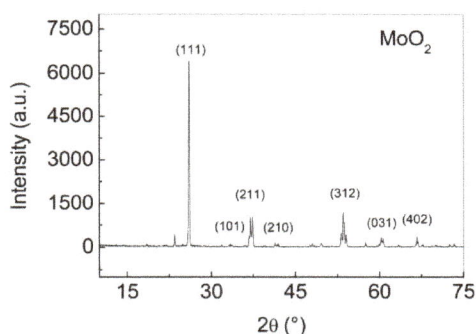

Figure 7. X-ray diffraction (XRD) of MoS$_2$ powder after annealing in Ar/H$_2$ and H$_2$O vapor at 1000 °C for 12 h. The peak positions are indexed to monoclinic MoO$_2$ (JCPDS NO. 00-032-0671).

4. Conclusions

In summary, we have successfully prepared monolayer/multi-layer MoS$_2$ through a H$_2$O vapor-modified vapor deposition method on a SiO$_2$/Si substrate. The growth of MoS$_2$ is highly sensitive to the presence of H$_2$O. The results reveal that H$_2$O increases the nucleation density of MoS$_2$ flakes. The Raman, PL, and AFM revealed that both monolayer and multi-layer MoS$_2$ were formed. Under extended water exposure, a continuous MoS$_2$ film was formed. Using XRD, we showed that MoO$_2$ was formed by the reaction between MoS$_2$ and water, which resulted in the observed enhancement in the nucleation and growth.

Supplementary Materials: The following are available online at http://www.mdpi.com/2079-6412/8/2/78/s1, Figure S1: Raman spectra of a multi-layers MoS$_2$ growth on the SiO2/Si substrate tested at different locations, corresponding to the data in Figure 3a–d; Figure S2: AFM image of monolayer MoS$_2$ showing the edge of the domain. Scale bar represents 1 μm in AFM image; Table S1: Average height of the monolayer MoS$_2$ corresponding to the data in Figure S2 and Figure 6.

Acknowledgments: This work was supported by the Natural Science Foundation of Zhejiang Province, China Projects (LY16E020008) and Chinese NSF Projects (61106100). We thank Haitao Liu (Department of Chemistry, University of Pittsburgh, USA) for his kind assistance with data analysis and paper writing.

Author Contributions: Shichao Zhao conceived and designed the experiments and wrote the paper; Jiaxin Weng, Shengzhong Jin and Yanfei Lv performed the experiments; Shichao Zhao and Zhenguo Ji analyzed the data.

Conflicts of Interest: The authors declare no conflict of interest.

References

1. Buscema, M.; Barkelid, M.; Zwiller, V.; van der Zant, H.S.J.; Steele, G.A.; Castellanos-Gomez, A. Large and tunable photothermoelectric effect in single-layer MoS$_2$. *Nano Lett.* **2013**, *13*, 358–363. [CrossRef] [PubMed]
2. Li, X.X.; Fan, Z.Q.; Liu, P.Z.; Chen, M.L.; Liu, X.; Jia, C.K.; Sun, D.M.; Jiang, X.W.; Han, Z.; Bouchiat, V.; et al. Gate-controlled reversible rectifying behaviour in tunnel contacted atomically-thin MoS$_2$ transistor. *Nat. Commun.* **2017**, *8*, 970. [CrossRef] [PubMed]
3. Naylor, C.H.; Kybert, N.J.; Schneier, C.; Xi, J.; Romero, G.; Saven, J.G.; Liu, R.Y.; Johnson, A.T.C. Scalable production of molybdenum disulfide based biosensors. *ACS Nano* **2016**, *10*, 6173–6179. [CrossRef] [PubMed]
4. Shavanova, K.; Bakakina, Y.; Burkova, I.; Shtepliuk, I.; Viter, R.; Ubelis, A.; Beni, V.; Starodub, N.; Yakimova, R.; Khranovskyy, V. Application of 2D non-graphene materials and 2D oxide nanostructures for biosensing technology. *Sensors* **2016**, *16*, 223. [CrossRef] [PubMed]
5. Kalantar-zadeh, K.; Ou, J.Z. Biosensors based on two-dimensional MoS$_2$. *ACS Sens.* **2016**, *1*, 5–16. [CrossRef]
6. Lopez-Sanchez, O.; Lembke, D.; Kayci, M.; Radenovic, A.; Kis, A. Ultrasensitive photodetectors based on monolayer MoS$_2$. *Nat. Nanotechnol.* **2013**, *8*, 497–501. [CrossRef] [PubMed]
7. Tsai, M.L.; Su, S.H.; Chang, J.K.; Tsai, D.S.; Chen, C.H.; Wu, C.I.; Li, L.J.; Chen, L.J.; He, J.H. Monolayer MoS$_2$ heterojunction solar cells. *ACS Nano* **2014**, *8*, 8317–8322. [CrossRef] [PubMed]
8. Robertson, J.; Liu, X.; Yue, C.L.; Escarra, M.; Wei, J. Wafer-scale synthesis of monolayer and few-layer MoS$_2$ via thermal vapor sulfurization. *2D Mater.* **2017**, *4*, 045007. [CrossRef]
9. Kim, Y.; Bark, H.; Ryu, G.H.; Lee, Z.; Lee, C. Wafer-scale monolayer MoS$_2$ grown by chemical vapor deposition using a reaction of MoO$_3$ and H$_2$S. *J. Phys. Condens. Matter* **2016**, *28*, 184002. [CrossRef] [PubMed]
10. Yu, Y.F.; Li, C.; Liu, Y.; Su, L.Q.; Zhang, Y.; Cao, L.Y. Controlled scalable synthesis of uniform, high-quality monolayer and few-layer MoS$_2$ films. *Sci. Rep.* **2013**, *3*, 1866. [CrossRef] [PubMed]
11. Lee, Y.H.; Zhang, X.Q.; Zhang, W.J.; Chang, M.T.; Lin, C.T.; Chang, K.D.; Yu, Y.C.; Wang, J.T.W.; Chang, C.S.; Li, L.J.; et al. Synthesis of large-area MoS$_2$ atomic layers with chemical vapor deposition. *Adv. Mater.* **2012**, *24*, 2320–2325. [CrossRef] [PubMed]
12. Liang, T.; Xie, S.; Huang, Z.T.; Fu, W.F.; Cai, Y.; Yang, X.; Chen, H.Z.; Ma, X.Y.; Iwai, H.; Fujita, D.; et al. Elucidation of zero-dimensional to two-dimensional growth transition in MoS$_2$ chemical vapor deposition synthesis. *Adv. Mater. Interfaces* **2017**, *4*, 1600687. [CrossRef]
13. Bampoulis, P.; van Bremen, R.; Yao, Q.R.; Poelsema, B.; Zandvliet, H.J.W.; Sotthewes, K. Defect dominated charge transport and fermi level pinning in MoS$_2$/metal contacts. *ACS Appl. Mater. Interfaces* **2017**, *9*, 19278–19286. [CrossRef] [PubMed]
14. McDonnell, S.; Addou, R.; Buie, C.; Wallace, R.M.; Hinkle, C.L. Defect-dominated doping and contact resistance in MoS$_2$. *ACS Nano* **2014**, *8*, 2880–2888. [CrossRef] [PubMed]
15. Zhou, W.; Zou, X.L.; Najmaei, S.; Liu, Z.; Shi, Y.M.; Kong, J.; Lou, J.; Ajayan, P.M.; Yakobson, B.I.; Idrobo, J.C. Intrinsic structural defects in monolayer molybdenum disulfide. *Nano Lett.* **2013**, *13*, 2615–2622. [CrossRef] [PubMed]
16. Addou, R.; Colombo, L.; Wallace, R.M. Surface defects on natural MoS$_2$. *ACS Appl. Mater. Interfaces* **2015**, *7*, 11921–11929. [CrossRef] [PubMed]
17. Wang, S.S.; Rong, Y.M.; Fan, Y.; Pacios, M.; Bhaskaran, H.; He, K.; Warner, J.H. Shape evolution of monolayer MoS$_2$ crystals grown by chemical vapor deposition. *Chem. Mater.* **2014**, *26*, 6371–6379. [CrossRef]
18. Yang, S.Y.; Shim, G.W.; Seo, S.B.; Choi, S.Y. Effective shape-controlled growth of monolayer MoS$_2$ flakes by powder-based chemical vapor deposition. *Nano Res.* **2017**, *10*, 255–262. [CrossRef]
19. Wu, S.F.; Huang, C.M.; Aivazian, G.; Ross, J.S.; Cobden, D.H.; Xu, X.D. Vapor-solid growth of high optical quality MoS$_2$ monolayers with near-unity valley polarization. *ACS Nano* **2013**, *7*, 2768–2772. [CrossRef] [PubMed]
20. Jin, S.Z.; Zhao, S.C.; Weng, J.X.; Lv, Y.F. Mn-promoted growth and photoluminescence of molybdenum disulphide monolayer. *Coatings* **2017**, *7*, 78. [CrossRef]

21. Pisoni, A.; Jacimovic, J.; Barisic, O.S.; Walter, A.; Nafradi, B.; Bugnon, P.; Magrez, A.; Berger, H.; Revay, Z.; Forro, L. The role of transport agents in MoS_2 single crystals. *J. Phys. Chem. C* **2015**, *119*, 3918–3922. [CrossRef]

22. Castellanos-Gomez, A.; Barkelid, M.; Goossens, A.M.; Calado, V.E.; van der Zant, H.S.J.; Steele, G.A. Laser-thinning of MoS_2: On demand generation of a single-layer semiconductor. *Nano Lett.* **2012**, *12*, 3187–3192. [CrossRef] [PubMed]

23. Mak, K.F.; Lee, C.; Hone, J.; Shan, J.; Heinz, T.F. Atomically thin MoS_2: A new direct-gap semiconductor. *Phys. Rev. Lett.* **2010**, *105*, 136805. [CrossRef] [PubMed]

24. Splendiani, A.; Sun, L.; Zhang, Y.B.; Li, T.S.; Kim, J.; Chim, C.Y.; Galli, G.; Wang, F. Emerging photoluminescence in monolayer MoS_2. *Nano Lett.* **2010**, *10*, 1271–1275. [CrossRef] [PubMed]

25. Bampoulis, P.; Teernstra, V.J.; Lohse, D.; Zandvliet, H.J.W.; Poelsema, B. Hydrophobic ice confined between graphene and MoS_2. *J. Phys. Chem. C* **2016**, *120*, 27079–27084. [CrossRef]

26. Varghese, J.O.; Agbo, P.; Sutherland, A.M.; Brar, V.W.; Rossman, G.R.; Gray, H.B.; Heath, J.R. The influence of water on the optical properties of single-layer molybdenum disulfide. *Adv. Mater.* **2015**, *27*, 2734–2740. [CrossRef] [PubMed]

27. Kwiecinski, W.; Sotthewes, K.; Poelsema, B.; Zandvliet, H.J.W.; Bampoulis, P. Chemical vapor deposition growth of bilayer graphene in between molybdenum disulfide sheets. *J. Colloid Interface Sci.* **2017**, *505*, 776–782. [CrossRef] [PubMed]

28. Sim, H.; Lee, J.; Park, B.; Kim, S.J.; Kang, S.; Ryu, W.; Jun, S.C. High-concentration dispersions of exfoliated MoS_2 sheets stabilized by freeze-dried silk fibroin powder. *Nano Res.* **2016**, *9*, 1709–1722. [CrossRef]

29. Frindt, R.F. Single crystals of MoS_2 several molecular layers thick. *J. Appl. Phys.* **1966**, *37*, 1928–1929. [CrossRef]

30. Li, H.; Zhang, Q.; Yap, C.C.R.; Tay, B.K.; Edwin, T.H.T.; Olivier, A.; Baillargeat, D. From bulk to monolayer MoS_2: Evolution of Raman scattering. *Adv. Funct. Mater.* **2012**, *22*, 1385–1390. [CrossRef]

31. Zhao, S.C.; Surwade, S.P.; Li, Z.T.; Liu, H.T. Photochemical oxidation of CVD-grown single layer graphene. *Nanotechnology* **2012**, *23*, 355703. [CrossRef] [PubMed]

32. Blanco, E.; Sohn, H.Y.; Han, G.; Hakobyan, K.Y. The kinetics of oxidation of molybdenite concentrate by water vapor. *Metall. Mater. Trans. B-Process Metall. Mater. Process. Sci.* **2007**, *38*, 689–693. [CrossRef]

coatings

MDPI

Article

Evaluation of the Scaffolding Effect of Pt Nanowires Supported on Reduced Graphene Oxide in PEMFC Electrodes

Peter Mardle, Oliver Fernihough and Shangfeng Du *

School of Chemical Engineering, University of Birmingham, Edgbaston, Birmingham B15 2TT, UK;
PJM556@student.bham.ac.uk (P.M.); OXF537@student.bham.ac.uk (O.F.)
* Correspondence: s.du@bham.ac.uk; Tel.: +44-121-415-8696

Received: 1 January 2018; Accepted: 22 January 2018; Published: 25 January 2018

Abstract: The stacking and overlapping effect of two-dimensional (2D) graphene nanosheets in the catalyst coating layer is a big challenge for their practical application in proton exchange membrane fuel cells (PEMFCs). These effects hinder the effective transfer of reactant gases to reach the active catalytic sites on catalysts supported on the graphene surface and the removal of the produced water, finally leading to large mass transfer resistances in practical electrodes and poor power performance. In this work, we evaluate the catalytic power performance of aligned Pt nanowires grown on reduced graphene oxide (rGO) (PtNW/rGO) as cathodes in 16-cm^2 single PEMFCs. The results are compared to Pt nanoparticles deposited on rGO (Pt/rGO) and commercial Pt/C nanoparticle catalysts. It is found that the scaffolding effect from the aligned Pt nanowire structure reduces the mass transfer resistance in rGO-based catalyst electrodes, and a nearly double power performance is achieved as compared with the Pt/rGO electrodes. However, although a higher mass activity was observed for PtNW/rGO in membrane electrode assembly (MEA) measurement, the power performance obtained at a large current density region is still lower than the Pt/C in PEMFCs because of the stacking effect of rGO.

Keywords: PEMFC; nanowire; graphene; PtPd; 2D

1. Introduction

With the commercial release of the Toyota Mirai and Hyundai ix35 fuel cell vehicles, the viability of using proton-exchange membrane fuel cells (PEMFCs) as the power source in transport applications has been demonstrated. However, for the widespread adoption of such vehicles, larger power outputs and long-term durability are required at a reduced cost [1,2]. Due to the sluggish oxygen reduction reaction (ORR) at the cathode in PEMFCs, catalysis is required. Platinum group metals (PGM), and specifically platinum (Pt), are inherently the most active catalysts towards the ORR [3] as well as the most stable in the acidic conditions of the PEMFC. However, due to their high cost and low performance compared to theoretical values, much effort has been directed to developing novel methods of reducing the PGM content while increasing the catalytic activity [4].

One-dimensional (1D) Pt nanostructures such as nanowires and nanotubes have received increasing interest in recent years [5]. Sun et al. synthesized Pt nanowires supported on carbon black (PtNW/C) by the simple wet chemical route of using formic acid to reduce chloroplatinic acid to Pt metal [6]. This catalyst demonstrated enhanced ORR activities as the favored Pt growth along the (111) crystallographic plane promoted by the slow reduction rate, and the regular arrays improved the intrinsic catalytic activity and mass transport limitations, respectively. This theory has been supported by Du's group, in whose work the synthesis conditions were optimized for the in situ growth of PtNWs directly on carbon paper gas diffusion layers (GDLs) [7,8].

To help maximize catalyst utilization ratio in PEMFC electrodes as well as improve stability, PGM catalysts are commonly supported on carbon blacks [9]. High surface area carbon blacks such as the Vulcan XC-72R are typically used due to the large catalyst electrochemically active surface areas (ECSA) obtainable from the provided porous network. However, a caveat of the porous network is the fact that many PGM catalyst particles can become trapped inside nanopores, where electrolyte ionomer cannot access them to form effective active sites. Carbon blacks also suffer from carbon corrosion due to the relatively low oxidation potential of carbon [10]. In an attempt to address these issues as well as in order to minimize charge transfer resistances in the catalyst layer, various carbonaceous supports such as carbon nanotubes (CNTs), carbon nanofibers (CNFs) and graphene have been investigated [11–13]. Of particular interest is the reduced graphene oxide (rGO), wherein the impurities from incomplete reduction provide much needed nucleation sites on the otherwise pristine graphene sheets. These functional groups have also been shown to help improve the carbon monoxide poisoning tolerance of the impurity in fuel cell gases as well as enhance reaction activities [14–16].

In combining the two areas of interest, PtNWs have been grown on rGO supports [14] as well as sulfur-doped graphene (SG) [17]. Although the latter referenced work showed an even distribution of nanowires on the rGO supports, branched nanowires formed to the anchoring oxygen functional groups. To help address this issue, Du et al. [15] introduced palladium (Pd) nanoseeds onto PVP-functionalized rGO sheets to grow uniform PtNWs. With improved distribution on the entire rGO surface, uniform nanowires were demonstrated to lead to enhanced catalytic activity.

However, the two-dimensional (2D) nature of the graphene-based catalysts makes them very difficult to use in practical electrodes in PEMFCs. In the drying process of the catalyst layer made by coating rGO-based catalyst inks, the rGO nanosheets stack on top of each other to form a dense catalyst layer, exacerbating the mass transfer limitations for both the reactant fuel and the produced water diffusing through the layer [16,18]. Considering the encouraging ORR catalytic activities of PtNWs/rGO, the intention of this work is to evaluate the scaffolding effect of the unique PtNWs/rGO structure by in situ testing in single PEMFCs. Fuel cells with cathodes from Pt nanoparticles supported on rGO (Pt/rGO) and commercial Pt/C nanoparticle catalysts are also fabricated as benchmarks to help understand the stacking effect of rGO nanosheets and the enhanced scaffolding effect from the nanowire geometry.

2. Materials and Methods

2.1. Materials

Poly(N-vinyl-2-pyrrolidine) (PVP-K30, molecular weight = 30,000–40,000), H_2PtCl_6 (8 wt % in H_2O), $PdCl_2$, formic acid HCOOH, Isopropyl alcohol (IPA) and L-Ascorbic acid were purchased from Sigma-Aldrich (Dorset, UK). The $PdCl_2$ was dissolved in water to provide a solution of concentration 0.125 g mL^{-1}. All other chemicals were used as received. The single-layer graphene oxide (GO) dispersion (thickness 0.43–1.23 nm, diameter 1.5–5.5 μm, dispersed in water at 2 wt %) was purchased from U.S. Research Nanomaterials, Inc. (Houston, TX, USA). Johnson–Matthey (JM) electrocatalysts with 20 wt % Pt supported on carbon black (Pt/C 20 wt %) was purchased from Alfa Aesar (Lancashire, UK). The water used throughout was purified with a Millipore (Hertfordshire, UK) system.

2.2. Preparation of PVP-Functionalized rGO

Two and a half milliliters of the 2 wt % GO suspension were diluted with 197.5 mL of H_2O. The resulting dispersion was sonicated with a 130 W, 20 kHz sonic horn at 40% amplitude for 30 min. Then 800 mg of PVP were added and the mixture was stirred for 12 h. Next 200 mg of L-Ascorbic acid were added to the mixture, which was subsequently heated to 95 °C. The mixture was then stirred for 1 h before centrifugation once with H_2O. The rGO was then dispersed in H_2O to provide a 25 mL suspension. Two and a half milliliters of the as-prepared rGO suspension were diluted with 17.5 mL of H_2O. Two separate suspensions were prepared.

2.3. Preparation of Pt/rGO

The preparation of Pt/rGO followed the process reported [19]. Two milliliters of the H_2PtCl_6 solution were added to one of the diluted rGO suspensions. Separately, 62 mg of $NaBH_4$ were dissolved in 20 mL of H_2O, which was added dropwise to the reaction mixture under sonication by a sonic bath. After 1 h of sonication time, the reaction mixture was left to settle overnight before being washed and collected by centrifugation in H_2O twice, followed by once with IPA. The catalyst was dispersed in IPA to provide a suspension of concentration 5 mg_{Pt} mL^{-1}.

2.4. Preparation of Pt(NW)Pd/rGO

Pt(NW)Pd/rGO was prepared following the procedure previously reported by our group [15]. Typically, 53.3 µL of the stock $PdCl_2$ solution was added to one of the diluted suspensions of rGO. Then 0.8 mL of HCOOH was added and the mixture was left to stir at room temperature for 16 h. Afterwards, a further 40 mL of H_2O were added, followed by 2 mL of the H_2PtCl_6 solution. Next 6.64 mL of HCOOH were added and the reaction mixture was left to react for 72 h. Centrifugation and dispersion of this catalyst in IPA followed the same procedure detailed for Pt/rGO.

2.5. Membrane Electrode Assembly (MEA) Preparation and Physical Characterization

To obtain suitable catalyst loadings of 0.4 and 0.6 mg_{Pt} cm^{-2} within electrodes, sufficient quantities of commercial Pt/C, Pt/rGO and Pt(NW)Pd/rGO suspensions were added to separate sample vials and 113 µL of 10 wt % Nafion solution were added to each. In the case of Pt/C, 0.875 mL of IPA was added to disperse the catalyst black. The catalyst inks were sonicated by a 130 W, 20 kHz sonication horn for 10 min at 20% power. The gas diffusion electrodes (GDEs) were made by coating the catalyst inks through painting onto 16 cm^2 Sigracet 35 BC carbon paper gas diffusion layers (GDLs). After leaving them to dry at room temperature, the electrodes were hot-pressed to a 36 cm^2 piece of Nafion® 212 membrane at 1800 lb load for 2 min at 135 °C with commercial JM Pt/C anodes with a catalyst loading 0.4 mg_{Pt} cm^{-2}. The GDE surface was imaged using a Philips (Amsterdam, The Netherlands) XL-30 FEG Environmental SEM operating at 20 kV.

2.6. Single Cell Testing

The MEAs were tested in a PEM fuel cell stand (FCT-50S, PaxiTech-BioLogic, Grenoble, France) with electrochemical impedance spectroscopy (EIS) capabilities. A polytetrafluoroethylene (PTFE) gasket of thickness 254 µm was used at both the cathode and anode sides. The membrane was hydrated by holding the cell potential at 0.6 V for 14 h at the cell temperature of 80 °C. The reactant gases were fully humidified H_2/air at the anode and cathode, respectively, with a backpressure of 0.5/0.5 bar and stoichiometry of 1.3:2.4, respectively. To measure the electrochemically effective surface area (ECSA) of the catalysts in the electrodes, the cathode gas was then switched to N_2. After 30 min, 200 cycles of cyclic voltammetry (CV) scan were conducted to the cathodes between 0.05 and 1.2 V at a sweep rate of 50 mV s^{-1}, followed by three scans at 20 mV s^{-1}. The third scan at 20 mV s^{-1} was used for the ECSA calculation. The cell potential was then held at 0.5 V for 30 min and the average current obtained was used for H_2 crossover correction in the mass activity measurement in the MEA. The reactant gas at the cathode was then changed back to air and, after being held stable for 30 min, the cell polarization curve was recorded at 1 mV s^{-1} between the open circuit voltage (OCV) and ca. 0.3 V. Electrochemical impedance spectroscopy (EIS) analysis was run at 0.65 V in the frequency range 10 kHz–0.1 Hz with an amplitude of 10 mV. To measure the mass activity of the catalyst in the electrode, the cathode was then switched to O_2 gas and a polarization curve was recorded at 1 mV s^{-1} with fully humidified H_2/O_2 at stoichiometry of 2/9.5, respectively.

3. Results

3.1. SEM Analysis of GDEs

Figure 1 shows the surface SEM images of the GDEs made from Pt/C, Pt/rGO, and Pt(NW)Pd/rGO. The GDE with Pt/C nanoparticle catalysts displays many large cracks (Figure 1a) with a land size of several hundred micrometers, which is in agreement with our former research; these cracks are considered essential for gas diffusion through the whole catalyst layer, together with the mesoporous network formed by the high surface area carbon support (Figure 1b) [20]. When rGO is introduced, as shown in Figure 1c,e, the number of cracks significantly decreases. Both GDEs from Pt/rGO and Pt(NW)Pd/rGO exhibit a very similar structure. This change can be ascribed to the stacking of the rGO within the catalyst layer and the 2D rGO nanosheets overlapping with each other to form a much denser coating layer. Comparing with the one from Pt/C nanoparticles shown in Figure 1a, this dense structure provides much less of a path for the gas diffusion through the catalyst layer, and thus a larger mass transfer resistance and a lower power performance is expected in the operation of PEMFCs. Figure 1d,f also show very few large aggregates on the electrode surface, demonstrating the good distribution of the rGO-based catalysts within the catalyst layer.

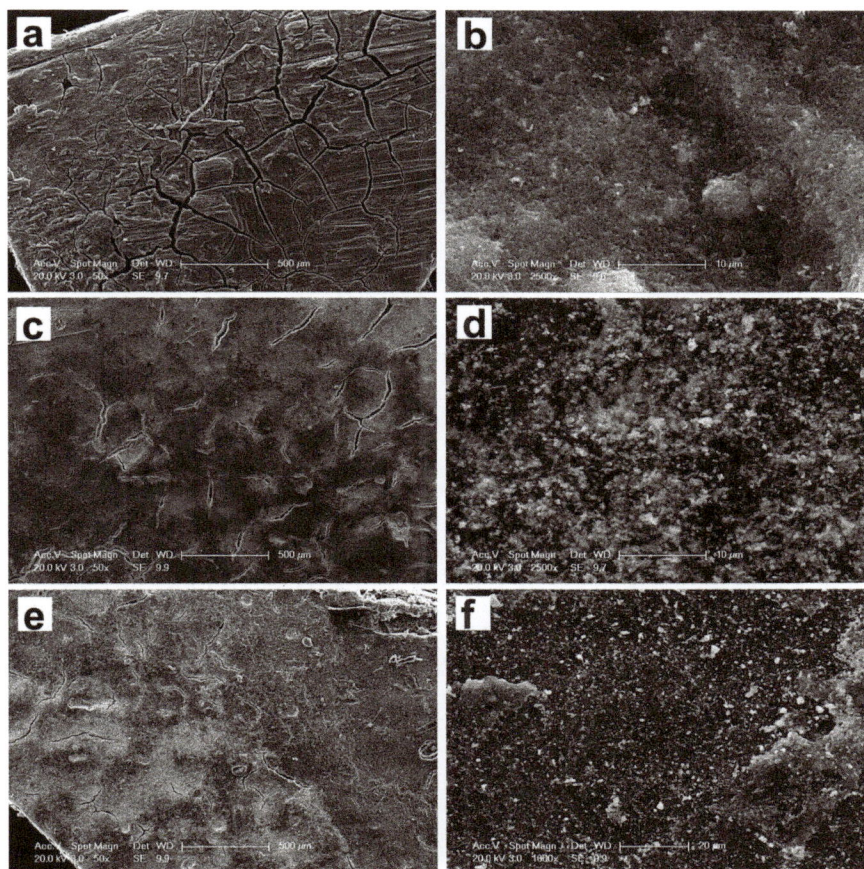

Figure 1. Surface SEM images of gas diffusion electrodes from (**a,b**) Pt/C 20% (JM); (**c,d**) Pt/rGO and (**e,f**) Pt(NW)Pd/rGO at a catalyst loading of 0.4 mg$_{Pt}$ cm^{-2}.

3.2. In Situ Testing

The GDEs were tested in H_2/air PEMFCs and the polarization and power density curves are shown in Figure 2. Pt(NW)Pd/rGO shows a much higher power performance than Pt/rGO. At 0.6 V, the conventional operation voltage for PEMFCs in practical applications, the power densities for the Pt(NW)Pd/rGO and Pt/rGO electrodes are 0.206 and 0.128 W cm^{-2}, respectively, at a catalyst loading of 0.4 mg_{Pt} cm^{-2}. The high power density of the Pt(NW)Pd/rGO electrode demonstrates the positive scaffolding effect from using nanowires. When increasing the catalyst loading from 0.4 to 0.6 mg_{Pt} cm^{-2}, the power performance for the Pt(NW)Pd/rGO and Pt/rGO electrodes increases very little—by only 9.7% or 3.1% to 0.226 or 0.132 W cm^{-2}, respectively—indicating the mass transfer limitations within the practical rGO-based electrodes, as mentioned by Antolini [16]. The increase of catalytic activities of the catalysts within the electrodes cannot really convert to the improvement of electrode power performance in PEMFC devices. The enhancement observed for nanowire electrodes even leads to high power density over the Pt/C nanoparticle electrode at a low current density range below 0.1 A cm^{-2}. However, the power performance is still much lower as compared with the Pt/C nanoparticle electrode at the large current density region, where a power density of 0.314 W cm^{-2} is achieved at 0.6 V. The poor power performance with rGO-based electrodes is in line with the dense electrode structure shown in the SEM analysis in Figure 1.

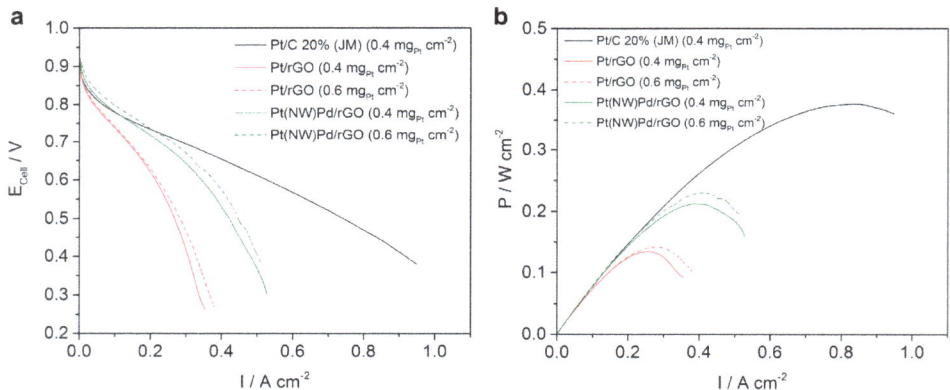

Figure 2. (**a**) Polarization and (**b**) power density curves obtained at a sweep rate of 1 mV s^{-1} at the cell temperature of 80 °C. The reactant gases are fully humidified H_2 at the anode and air at the cathode with backpressure of 0.5/0.5 bars and stoichiometry of 1.3/2.4, respectively.

To further understand the mechanisms behind the obtained performances, detailed analyses were conducted to obtain the ECSA, mass activity, and EIS of the involved electrodes (Figure 3); the results are summarized in Table 1. An acceptable ECSA value of 19.84 m^2 g^{-1} is obtained for the Pt/C electrode, which is ca. 1/3 of the value obtained in the liquid half-cell rotating disk electrode (RDE) measurement (ca. 50–70 m^2 g^{-1}) and in line with values reported in the literature [15,19]. However, the ECSA values observed for rGO-based electrodes are extremely low. By RDE measurement, the ECSA values obtained for Pt/rGO are usually similar to or even higher than Pt/C nanoparticle catalysts, but the value obtained here in the MEA is only 2.65 m^2 g^{-1}, which is much lower than that of Pt/C in the electrode. This lower ECSA for Pt/rGO further indicates a very low catalyst utilization ratio in the electrode (defined by the ratio of the ECSA observed in the electrode to the value by RDE measurement), ca. 5.45%, taking into account the ECSA value of 48.62 m^2 g^{-1} reported by the RDE measurement [19], resulting from the stacking and overlapping of the rGO nanosheets in the catalyst layer. For the Pt(NW)Pd/rGO, the ECSA value drops from 19.07 m^2 g^{-1} by RDE measurement [15] to 2.97 m^2 g^{-1} here. Although the utilization ratio is still low, ca. 15.59%, it has been much higher than

Pt/rGO, resulting from the scaffolding effect of the nanowires. However, the utilization ratio is still much lower as compared with the Pt/C nanoparticles. In addition, an increase of the catalyst loading from 0.4 to 0.6 mg_{Pt} cm^{-2} leads to slightly reduced ECSA values, resulting from the mass transfer limitation mentioned above. The mass activity in the MEA was measured under pure O_2 rather than air to reduce the influence of the mass transfer resistance in the catalyst layer [1,7]. The standard measurement protocol defined by the U.S. Department of Energy (DOE) was followed [1]. However, this test under oxygen still does not work very well for the Pt/rGO electrode, and a much lower mass activity of 0.016 A mg_{Pt}^{-1} is observed as compared with 0.041 A mg_{Pt}^{-1} for Pt/C. A similar value is also obtained for the catalyst loading at 0.6 mg cm^{-2}. For Pt(NW)Pd/rGO, this value reaches 0.044 A mg_{Pt}^{-1}, even higher than Pt/C. The higher mass activity of this nanowire structure in the electrode can be ascribed to two factors: (i) the scaffolding effect, which partially mitigates the severe stacking of the 2D rGO nanosheets, thus improving the mass transfer performance; and (ii) the excellent specific catalytic activity from the special surface properties of the unique nanowires [6,8]. The improvement of the nanowire scaffolding effect and the mass transfer limitation of rGO-based electrodes are further confirmed in Figure 3b. The mass transfer impedance of Pt(NW)Pd/rGO is much lower than Pt/rGO, but still higher than Pt/C in the MEAs, as shown in Figure 3c.

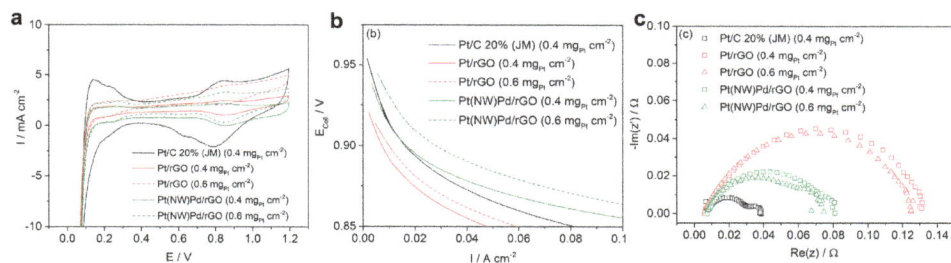

Figure 3. (**a**) Cathode cyclic voltammetry (CV) scan between 0.05–1.2 V at a sweep rate of 20 mV s^{-1} with a cell temperature of 80 °C. The reactant gases are fully humidified H_2 at the anode and N_2 at the cathode with backpressure of 0.5/0.5 bars and stoichiometry of 1.3/2.4, respectively. (**b**) Resistance and H_2 crossover corrected polarization curves obtained at a sweep rate of 1 mV s^{-1} at the cell temperature of 80 °C. The reactant gases are fully humidified H_2 at the anode and O_2 at the cathode with backpressure of 0.5/0.5 bars and stoichiometry of 2/9.5, respectively. (**c**) EIS spectra at 0.65 V with amplitude 10 mV in the frequency range 10 kHz–0.1 Hz for cells with air as the reactant gas (the same settings as for Figure 2a).

Table 1. ECSA, specific (S.A.) and mass activities (M.A.) of the catalysts in membrane electrode assembly in single cells (derived from Figure 3a,b).

Cathode	Power Density (0.6 V) [W cm^{-2}]	ECSA [m^2 g_{Pt}^{-1}]	$i_{0.9V}$ [mA]	S.A.$_{-0.9V}$ [μA cm^{-2}]	M.A.$_{-0.9V}$ [A mg_{Pt}^{-1}]
Pt/C 20% (JM) (0.4 mg_{Pt} cm^{-2})	0.314	19.84	263.520	208	0.041
Pt/rGO (0.4 mg_{Pt} cm^{-2})	0.128	2.65	104.480	616	0.016
Pt/rGO (0.6 mg_{Pt} cm^{-2})	0.132	2.22	146.720	689	0.015
Pt(NW)Pd/rGO (0.4 mg_{Pt} cm^{-2})	0.206	2.97	278.560	1466	0.044
Pt(NW)Pd/rGO (0.6 mg_{Pt} cm^{-2})	0.226	2.71	453.920	1744	0.047

4. Discussion

In the fabrication of PEMFC electrodes, the catalyst ink is coated onto the GDL or membrane surface. In the drying process, the organic solvent (e.g., IPA) in the catalyst ink is evaporated, leaving a porous catalyst layer [1,20]. However, the 2D graphene nanosheets tend to stack on top of each other and overlap to form a dense structure within the catalyst layer, which blocks the diffusion of the reactant gas and the removal of the produced water in operating PEMFCs and causes poor power performance. Even by increasing the catalyst loading in the electrodes, an effective improvement of the power performance is not achieved because of the mass transfer limitation. Thus, the conclusion is usually drawn that 2D graphene-based materials are not suitable as catalyst supports for practical electrodes in PEMFC devices, despite the much better activities that were observed by the liquid half-cell RDE measurement, where the influence of the mass transfer can nearly be ignored [16].

Nanowires have a large aspect ratio. When they are grown on a 2D nanosheet surface, the unique geometry can form scaffold-like structures and thus partially release the stacking influence of the nanosheets. In fact, in graphene-related research, this function has been used by Si et al. [18] to exfoliate graphene sheets by the use of Pt nanoparticles. In their work the presence of Pt nanoparticles impregnated on the surface of graphene showed a surface area (obtained by BET measurements) around 20 times larger than the graphene without Pt. This scaffolding effect in the nanowire/rGO electrode similarly improves the mass transfer performance within the catalyst layer and an even higher power density is achieved than with Pt/C at a low current density range (Figure 2a). However, the rGO nanosheets usually have a much larger size (on the micrometer scale, as shown in Figure 1 and [15]) than the length of Pt nanowires (usually 20–200 nm), so, along with the difficulty of nanowires covering the entire surface of every rGO nanosheet, the stacking of rGO still cannot be fully avoided. Therefore, the improvement of this scaffolding effect is still limited. The stacking and overlapping effects caused by the large rGO nanosheets dominates the structures of the electrodes; even with nanowires, large cracks can still not form as with Pt/C nanoparticle catalysts, and quite similar electrode structures are observed in SEM for electrodes made of both Pt/rGO and Pt(NW)Pd/rGO (Figure 1). When the fuel cell is operated at a large current density, as shown in Figure 2a, the amount of reactant gas required for the reaction and the water produced increased dramatically; thus, the scaffolding space provided by the aligned nanowires between the rGO nanosheets is insufficient and thus poor power performance is observed again. This finally causes a clear crossover of the polarization curves of the two MEAs from Pt(NW)Pd/rGO and Pt/C, as shown in Figure 2a. This also leads to another suggestion that if a large enough scaffolding effect can be introduced to the electrode structure to avoid the stacking and overlapping to form similar large cracks as that of Pt/C, e.g., by a further increase of the nanowire length and the improvement of their distribution on the rGO surface, or by reducing all rGO nanosheets to a relative smaller size, or through investigating new alternative electrode preparation methods [21,22], high power performance is still possible for rGO-based electrodes for PEMFCs in real-life operation. Such pursuits are of particular importance when considering the improved resistance of nanowires to coarsening in comparison to Pt/C nanoparticle catalysts [23,24] and the enhanced resistance of rGO supports to carbon corrosion [25], both major limitations in the commercial viability of PEMFCs.

5. Conclusions

In this work, we evaluated the catalyst electrodes made from Pt(NW)Pd/rGO and Pt/rGO in PEMFCs and the results were compared to Pt/C nanoparticle catalysts. The results demonstrated that the introduction of rGO-based catalysts tended to form a dense structure within the catalyst layer that limited the mass transfer in fuel cell operation and finally resulted in poor power performance. The aligned nanowires grown on rGO surface can work as a scaffold, thus improving the mass transfer limitation. At 0.6 V, power densities of 0.206 and 0.128 W cm^{-2} were obtained for Pt(NW)Pd/rGO and Pt/rGO electrodes, respectively, at the catalyst loading of 0.4 mg$_{Pt}$ cm^{-2}. The mass activity for Pt(NW)Pd/rGO was 0.044 A mg$_{Pt}$$^{-1}$ in MEA, which is much higher than 0.016 A mg$_{Pt}$$^{-1}$ of the Pt/rGO, and even better than 0.041 A mg$_{Pt}$$^{-1}$ for Pt/C. However, the power density obtained for the nanowire/rGO electrode is still lower than Pt/C at 0.6 V. Nevertheless, this work demonstrates the possibility of using graphene-based materials as a catalyst support for fuel cell devices. To really bring this technology into practical application, further study is required to understand the complicated influences of the nanowire length, distribution on rGO, size and structure of rGO, and electrode preparation method.

Acknowledgments: This work was funded by the EPSRC Centre for Doctoral Training in Fuel Cells and their Fuels (EP/L015749/1).

Author Contributions: Shangfeng Du and Peter Mardle conceived and designed the experiments; Oliver Fernihough performed the experiments; Peter Mardle and Oliver Fernihough analyzed the data; Shangfeng Du and Peter Mardle wrote the paper.

Conflicts of Interest: The authors declare no conflict of interest. The founding sponsors had no role in the design of the study; in the collection, analyses, or interpretation of data; in the writing of the manuscript, and in the decision to publish the results.

References

1. Gasteiger, H.A.; Kocha, S.S.; Sompalli, B.; Wagner, F.T. Activity benchmarks and requirements for Pt, Pt-alloy, and non-Pt oxygen reduction catalysts for PEMFCs. *Appl. Catal. B Environ.* **2005**, *56*, 9–35. [CrossRef]
2. Office of Energy Efficiency and Renewable Energy. *Fuel Cell Technologies Office Multi-Year Research, Development, and Demonstration Plan. Section 3.4: Fuel Cells*; US Department of Energy: Washington, DC, USA, 2016.
3. Norskov, J.K.; Rossmeisl, J.; Logadottir, A.; Lindqvist, L.; Kitchin, J.R.; Bligaard, T.; Jonsson, H. Origin of the overpotential for oxygen reduction at a fuel-cell cathode. *J. Phys. Chem. B* **2004**, *108*, 17886–17892. [CrossRef]
4. Nie, Y.; Li, L.; Wei, Z.D. Recent advancements in Pt and Pt-free catalysts for oxygen reduction reaction. *Chem. Soc. Rev.* **2015**, *44*, 2168–2201. [CrossRef] [PubMed]
5. Lu, Y.X.; Du, S.F.; Steinberger-Wilckens, R. One-dimensional nanostructured electrocatalysts for polymer electrolyte membrane fuel cells—A review. *Appl. Catal. B* **2016**, *199*, 292–314. [CrossRef]
6. Sun, S.H.; Jaouen, F.; Dodelet, J.P. Controlled Growth of Pt Nanowires on Carbon Nanospheres and Their Enhanced Performance as Electrocatalysts in PEM Fuel Cells. *Adv. Mater* **2008**, *20*, 3900–3904. [CrossRef]
7. Lu, Y.X.; Du, S.F.; Steinberger-Wilckens, R. Temperature-controlled growth of single-crystal Pt nanowire arrays for high performance catalyst electrodes in polymer electrolyte fuel cells. *Appl. Catal. B* **2015**, *164*, 389–395. [CrossRef]
8. Du, S.F.; Lin, K.J.; Malladi, S.K.; Lu, Y.X.; Sun, S.H.; Xu, W.; Steinberger-Wilckens, R.; Dong, H.S. Plasma nitriding induced growth of Pt-nanowire arrays as high performance electrocatalysts for fuel cells. *Sci. Rep.* **2014**, *4*, 6439. [CrossRef] [PubMed]
9. Antolini, E. Carbon supports for low-temperature fuel cell catalysts. *Appl. Catal. B* **2009**, *88*, 1–24. [CrossRef]
10. Castanheira, L.; Dubau, L.; Mermoux, M.; Berthome, G.; Caque, N.; Rossinot, E.; Chatenet, M.; Maillard, F. Carbon Corrosion in Proton-Exchange Membrane Fuel Cells: From Model Experiments to Real-Life Operation in Membrane Electrode Assemblies. *ACS Catal.* **2014**, *4*, 2258–2267. [CrossRef]
11. Wildgoose, G.G.; Banks, C.E.; Compton, R.G. Metal nanoparticles and related materials supported on carbon nanotubes: Methods and applications. *Small* **2006**, *2*, 182–193. [CrossRef] [PubMed]
12. Zheng, J.S.; Wang, M.X.; Zhang, X.S.; Wu, Y.X.; Li, P.; Zhou, X.G.; Yuan, W.K. Platinum/carbon nanofiber nanocomposite synthesized by electrophoretic deposition as electrocatalyst for oxygen reduction. *J. Power Sources* **2008**, *175*, 211–216. [CrossRef]
13. Jafri, R.I.; Rajalakshmi, N.; Ramaprabhu, S. Nitrogen doped graphene nanoplatelets as catalyst support for oxygen reduction reaction in proton exchange membrane fuel cell. *J. Mater. Chem.* **2010**, *20*, 7114–7117. [CrossRef]
14. Luo, Z.M.; Yuwen, L.H.; Bao, B.Q.; Tian, J.; Zhu, X.R.; Weng, L.X.; Wang, L.H. One-pot, low-temperature synthesis of branched platinum nanowires/reduced graphene oxide (BPtNW/RGO) hybrids for fuel cells. *J. Mater. Chem.* **2012**, *22*, 7791–7796. [CrossRef]
15. Du, S.F.; Lu, Y.X.; Steinberger-Wilckens, R. PtPd nanowire arrays supported on reduced graphene oxide as advanced electrocatalysts for methanol oxidation. *Carbon* **2014**, *79*, 346–353. [CrossRef]
16. Antolini, E. Graphene as a new carbon support for low-temperature fuel cell catalysts. *Appl. Catal. B* **2012**, *123*, 52–68. [CrossRef]
17. Wang, R.Y.; Higgins, D.C.; Hoque, M.A.; Lee, D.; Hassan, F.; Chen, Z.W. Controlled Growth of Platinum Nanowire Arrays on Sulfur Doped Graphene as High Performance Electrocatalyst. *Sci. Rep.* **2013**, *3*, 2431–2437. [CrossRef] [PubMed]

18. Si, Y.C.; Samulski, E.T. Exfoliated Graphene Separated by Platinum Nanoparticles. *Chem. Mater.* **2008**, *20*, 6792–6797. [CrossRef]

19. Vu, T.H.T.; Tran, T.T.T.; Le, H.N.T.; Tran, L.T.; Nguyena, P.H.T.; Nguyen, M.D.; Quynh, B.H. Synthesis of Pt/rGO catalysts with two different reducing agents and their methanol electrooxidation activity. *Mater. Res. Bull.* **2016**, *73*, 197–203. [CrossRef]

20. Millington, B.; Du, S.F.; Pollet, B.G. The effect of materials on proton exchange membrane fuel cell electrode performance. *J. Power Sources* **2011**, *196*, 9013–9017. [CrossRef]

21. Úbeda, D.; Lobato, J.; Cañizares, P.; Pinar, F.J.; Zamora, H.; Fernández-Marchante, C.M.; Rodrigo, M.A. Using Current Distribution Measurements to Characterize the Behavior of HTPEMFCs. *Chem. Eng. Trans.* **2014**, *41*, 229–234. [CrossRef]

22. Mardle, P.; Du, S.F. Materials for PEMFC Electrodes. In *Reference Module in Materials Science and Materials Engineering*; Hashmi, S., Ed.; Elsevier: Amsterdam, The Netherland, 2017; pp. 1–13, ISBN 978-0-12-803581-8.

23. Sun, S.; Zhang, G.; Geng, D.; Chen, Y.; Cai, M.; Sun, X. A Highly Durable Platinum Nanocatalyst for Proton Exchange Membrane Fuel Cells: Multiarmed Starlike Nanowire Single Crystal. *Angew. Chem. Int. Ed.* **2011**, *50*, 422–426. [CrossRef] [PubMed]

24. Lu, Y.X.; Du, S.F.; Steinberger-Wilckens, R. Three-dimensional catalyst electrodes based on PtPd nanodendrites for oxygen reduction reaction in PEFC applications. *Appl. Catal. B* **2016**, *187*, 108–114. [CrossRef]

25. Sun, K.G.; Chung, J.S.; Hur, S.H. Durability Improvement of Pt/RGO Catalysts for PEMFC by Low-Temperature Self-Catalyzed Reduction. *Nanoscale Res. Lett.* **2015**, *10*, 257–263. [CrossRef] [PubMed]

coatings

MDPI

Communication

Enhanced Efficiency of Dye-Sensitized Solar Counter Electrodes Consisting of Two-Dimensional Nanostructural Molybdenum Disulfide Nanosheets Supported Pt Nanoparticles

Chao-Kuang Cheng [1], Jeng-Yu Lin [2], Kai-Chen Huang [3], Tsung-Kuang Yeh [1] and Chien-Kuo Hsieh [3],*

[1] Department of Engineering and System Science, National Tsing Hua University, Hsinchu 30013, Taiwan; ckcheng@gapp.nthu.edu.tw (C.-K.C.); tkyeh@mx.nthu.edu.tw (T.-K.Y.)
[2] Department of Chemical Engineering, Tatung University, Taipei 104, Taiwan; jylin@ttu.edu.tw
[3] Department of Materials Engineering, Ming Chi University of Technology, New Taipei City 24301, Taiwan; a0983156660@gmail.com
* Correspondence: jack_hsieh@mail.mcut.edu.tw; Tel.: +886-2-2908-9899 (ext. 4438)

Academic Editors: Federico Cesano and Domenica Scarano
Received: 5 September 2017; Accepted: 10 October 2017; Published: 13 October 2017

Abstract: This paper reports architecturally designed nanocomposites synthesized by hybridizing the two-dimensional (2D) nanostructure of molybdenum disulfide (MoS_2) nanosheet (NS)-supported Pt nanoparticles (PtNPs) as counter electrodes (CEs) for dye-sensitized solar cells (DSSCs). MoS_2 NSs were prepared using the hydrothermal method; PtNPs were subsequently reduced on the MoS_2 NSs via the water–ethylene method to form PtNPs/MoS_2 NSs hybrids. The nanostructures and chemical states of the PtNPs/MoS_2 NSs hybrids were characterized by high-resolution transmission electron microscopy and X-ray photoelectron spectroscopy. Detailed electrochemical characterizations by electrochemical impedance spectroscopy, cyclic voltammetry, and Tafel-polarization measurement demonstrated that the PtNPs/MoS_2 NSs exhibited excellent electrocatalytic activities, afforded a higher charge transfer rate, a decreased charge transfer resistance, and an improved exchange current density. The PtNPs/MoS_2 NSs hybrids not only provided the exposed layers of 2D MoS_2 NSs with a great deal of catalytically active sites, but also offered PtNPs anchored on the MoS_2 NSs enhanced I_3^- reduction. Accordingly, the DSSCs that incorporated PtNPs/MoS_2 NSs CE exhibited an outstanding photovoltaic conversion efficiency (PCE) of 7.52%, which was 8.7% higher than that of a device with conventional thermally-deposited platinum CE (PCE = 6.92%).

Keywords: MoS_2 nanosheets; Pt nanoparticles; counter electrode; dye-sensitized solar cells

1. Introduction

A typical dye-sensitized solar cell (DSSC) is usually fabricated by dye-loaded TiO_2 nanoparticles coated on transparent conductive glass to act as the working electrode (WE), a Pt film deposited on the transparent conductive glass as the counter electrode (CE), and an electrolyte containing an iodide/triiodide (I^-/I_3^-) redox couple between the WE and CE [1]. Recently, considerable efforts have focused on improving the energy conversion efficiency and long-term stability of the dye, electrolyte, and working electrode [2–7]. However, a high catalytic material for CE is extremely important for promoting the charge transfer rate for I_3^- reduction. Therefore, a CE with high electrochemical activity and low internal resistance is crucial to yielding high photovoltaic conversion efficiency (PCE).

Recently, inspired by the discovery of graphene—which opened up the new research field of two-dimensional (2D) nanomaterials [8,9]—studies on 2D nanomaterials have attracted great attention. As a typical transition metal dichalcogenide, 2D layered nanostructural molybdenum disulfide (MoS_2) is similar to the graphene structure, and is composed of three stacked atomic layers (a Mo layer sandwiched between two S layers, S-Mo-S) [10]. Layered nanostructural 2D MoS_2 has been extensively investigated as a promising catalyst in electrochemical applications in recent years such as H_2 evolution [11], Li-ion batteries [12], and DSSCs [13]. For instance, Jaramillo et al. reported electrochemical H_2 production with MoS_2 nanocatalysts based on their catalytically active sites for H_2 evolution [11]. Very recently, Mohammad et al. reported that the ultrathin MoS_2 nanostructured films possessed the outstanding catalytic performance of the CE for DSSCs, with rich catalytic active sites provided by the MoS_2 nanosheet crystal structure [13]. Furthermore, due to the high specific surface area of 2D nanomaterials, zero-dimensional (0D) catalytic nanoparticles anchored on the 2D nanomaterials to form 0D/2D functional hybrids bring novel properties that are different from those of their individual intrinsic properties. For example, graphene-supported Pt nanoparticles (PtNPs) such as PtNPs/graphene hybrids, have been investigated as advanced electrocatalysts for achieving superior activities of CEs in producing high-performing DSSCs [14–17]. Yanyan et al. reported that graphene acted as an ideal support for uniformly disperse PtNPs, which was intrinsically important for an effective CE in DSSCs [14]. Min-Hsin et al. reported that PtNPs/graphene CE displayed larger realistic electroactive surface areas and a constant higher intrinsic heterogeneous rate to improve the electrocatalytic abilities for the reduction of I_3^-, and therefore improved the PCE of DSSCs [15].

In terms of the above considerations, in this study, we report on our investigation of the performance of DSSCs that used the PtNPs/MoS_2 NSs hybrid nanoarchitecture as the CE. This hybrid CE, constituted by 2D layered MoS_2 NS-supported 0D PtNPs, showed a number of advantages. The MoS_2 NSs not only provided exposed-layer active sites for I_3^- reduction, but also provided large surface areas for PtNP anchoring on the MoS_2 NSs to enhance electrochemical activities. The DSSCs fabricated with the PtNPs/MoS_2 NSs CE had an increased exchange current density and reduced charge-transfer resistance, resulting in a superior PCE of 7.52%, 8.7% higher than that of a conventional thermally-deposited Pt CE (6.92%).

2. Materials and Methods

2.1. Preparation of MoS_2 NSs and PtNPs/MoS_2 NSs

Two steps were used to synthesize PtNPs/MoS_2 NSs CE. In Step 1, the hydrothermal method was applied to synthesize the MoS_2 NSs. Ammonium tetrathiomolybdate ((NH_4)$_2MoS_4$) powder (99.99% purity, ProChem Inc., Rockford, IL, USA) weighing 0.5 g was added to an aqueous solution with 5 mL of HCl in 100 mL deionized water. Subsequently, the aforementioned solution was transferred into a Teflon-lined autoclave and heated to 250 °C for 12 h. The suspension was then washed and centrifuged with deionized water and ethanol several times, respectively. The resulting powder was finally dried in vacuum at 60 °C to obtain MoS_2 NSs. In Step 2, PtNPs/MoS_2 NSs hybrids were prepared using a water–ethylene method [17]. MoS_2 NSs (50 mg) were ultrasonically dispersed in a mixture containing 30 mL of deionized water, 100 mL of ethylene glycol (EG), and 1.5 mL of 0.05 M H_2PtCl_6 aqueous solution. This mixture was then heated at 120 °C with stirring for 6 h. Subsequently, the composite was washed and centrifuged with deionized water and ethanol, respectively, for a total of six times. Finally, the resulting powder was dried in vacuum at 60 °C to obtain the PtNPs/MoS_2 NSs hybrids.

2.2. Fabrication of Various CEs and Assembly of DSSCs

Fluorine-doped tin oxide (FTO) transparent glasses (TEC-7, 2.2 mm, Hartford glass, Hartford, IN, USA) were used as the substrates for CEs and WEs. Prior to the fabrications of CEs and WEs, FTO glasses were ultrasonically cleaned sequentially in detergent, acetone (overnight), distilled water

(DI water, 1 h), and ethanol (1 h). The CEs with MoS$_2$ NSs and PtNPs/MoS$_2$ NSs hybrids were fabricated as follows: 5 mg MoS$_2$ NSs powder and PtNPs/MoS$_2$ NSs powder were added to 5 mL *N*,*N*-dimethylformamide (DMF) for dispersion and then sonicated for 1 h, respectively. Subsequently, the dispersed solutions were coated on FTO glasses by spin coating technology to control the flatness and thickness of the films. Finally, the prepared samples were dried in vacuum at 60 °C for 1 h to obtain the MoS$_2$ NSs CE and the PtNPs/MoS$_2$ NSs CE. In addition, a thermal deposition Pt (TD-Pt) CE was also prepared as a reference electrode by dropping a H$_2$PtCl$_6$ isopropanol solution on a FTO glass annealed at 450 °C for 20 min [18].

A screen-printing method was carried out to prepare the WEs: 10 μm TiO$_2$ nanoparticle films were coated onto the FTO glasses, and were then placed in the furnace for calcination at 550 °C in ambient air for 30 min. After slowly cooled to room temperature (RT), the TiO$_2$ WEs were removed from the furnace and immersed in a N719 (Solaronix) solution (3×10^{-3} M in a 1:1 volumetric mixture of acetonitrile and *tert*-butylalcohol) at RT for 24 h.

The DSSCs were assembled as follows. After the dye adsorption process, the dye-adsorbed TiO$_2$ WE was assembled with various CEs as the sandwich-type cell, and sealed with a 60 μm hot-melt surlyn (SX1170-60, Solaronix, Aubonne, Switzerland) between WE and CE. Then, the commercial iodide-based electrolyte (TDP-LE-M, Jintex Corporation Ltd., Taipei, Taiwan) was injected into the space between the two electrodes after the assembling process. A Class A quality solar simulator with a light intensity of 100 mW·cm^{-2} (AM 1.5) was used as the light source to illuminate the DSSCs devices to measure the photocurrent-voltage characteristics.

2.3. Characterizations

Transmission electron microscopy (TEM, JEM-2100F, JOEL, Tokyo, Japan) was used to examine the nanostructures of the prepared PtNPs/MoS$_2$ NSs hybrids. The sample for TEM was prepared by dropping the sample solution on a copper grid coated with a carbon film. X-ray photoelectron spectroscopy (XPS) was carried out using a PHI Quantera SXM/AES 650 (ULVAC-PHI Inc., Kanagawa, Japan) system with a hemispherical electron analyzer and a scanning monochromated Al Kα ($h\nu$ = 1486.6 eV) X-ray source to investigate the chemical states of Mo, S, and Pt. To study the chemical states of the PtNPs/MoS$_2$ NSs hybrids, XPSPEAK 4.1 was used for fitting the obtained curves, peak de-convolution and assignment of binding energies, and referenced to the adventitious C 1s peak at 284.6 eV. For spectrum analysis, the background signal was subtracted by Shirley's method, and curve fitting was performed by using a Gaussian-Lorentzian peak after Shirley background correction.

The catalytic abilities of the CEs were examined by cyclic voltammetry (CV) measurements, equipped with a three-electrode configuration using potentiostat/galvanostat (PGSTAT 302N, Metrohm Autolab, Eco Chemie, Utrecht, The Netherlands) in an acetonitrile-based solution consisting of 10 mM LiI, 1.0 mM I$_2$, and 0.1 M LiClO$_4$. The Pt wire and an Ag/AgNO$_3$ electrode were employed as the counter and reference electrodes, respectively. Electrochemical impedance spectroscopy (EIS) was carried out to study the electrochemical properties of the CEs. The aforementioned potentiostat/galvanostat—equipped with a frequency response analysis (FRA) module—was used for the EIS analyses in a frequency range between 10^6 Hz and 10^{-2} Hz. Tafel polarization curves were also measured using the potentiostat/galvanostat equipped with a linear polarization module to further investigate the catalytic activities at the electrolyte-electrode interface of various CEs. Both EIS and Tafel-polarization measurements were obtained using symmetrical devices in the dark. The photocurrent–voltage characteristics of DSSC devices were measured under simulated solar illumination (AM 1.5, 100 mW·cm^{-2}, Oriel 91160, Newport Corporation, Irvine, CA, USA), equipped with an AM 1.5 G filter (Oriel 81088A, Newport Corporation, Irvine, CA, USA) and a 300 W xenon lamp (Oriel 6258, Newport Corporation, Irvine, CA, USA). The simulated incident light intensity was calibrated using a reference Si cell (calibrated at NREL, PVM-81).

3. Results

3.1. Nanostructural Features and Composition

The morphologies and nanostructures were studied using TEM and HRTEM (JEM-2100F, JOEL, Tokyo, Japan). Figure 1a shows a typical TEM image of the PtNPs/MoS$_2$ NSs. As we can see from Figure 1a, the PtNPs were uniformly dispersed on the MoS$_2$ NSs without aggregation. The HRTEM image in Figure 1b shows the plentifully exposed plans with an interlayer distance of the MoS$_2$ of about 0.64 nm, corresponding to the spacing between (002) planes of MoS$_2$, similar to previous studies in Reference [12]. In addition, the HRTEM image clearly shows PtNPs with a uniform size on the MoS$_2$ NSs, and particle sizes of the PtNPs in the range of 3–5 nm. The inset in Figure 1b shows a Pt nanoparticle with the lattice distance of 0.22 nm corresponding to the (111) plane of crystalline Pt.

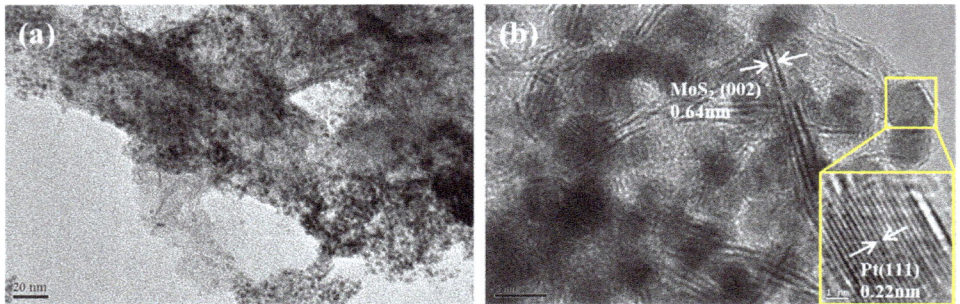

Figure 1. (**a**) TEM image of PtNPs distributed over MoS$_2$ NSs; (**b**) HRTEM of PtNPs/MoS$_2$ NSs, the inset showed the lattice distance of 0.22 nm corresponding to the (111) plane of Pt.

Figure 2 and Table 1 show the XPS fitting results of PtNPs/MoS$_2$ NSs. Figure 2a shows the XPS spectrum of the wide spectral region of the MoS$_2$ NSs and PtNPs/MoS$_2$ NSs, respectively. The relatively characteristic peaks of the elements are also illustrated in Figure 2a. As seen in Figure 2a, when compared with MoS$_2$ NSs, the Pt 4f peak of PtNPs/MoS$_2$ NSs can be clearly seen. Figure 2b–d show the chemical states of Pt 4f, Mo 3d, and the S 2p orbitals of the PtNPs/MoS$_2$ NSs, respectively. Figure 2b shows the high-resolution Pt 4f spectra, where the main peaks at 71.2 and 74.5 eV correspond to Pt^{0}4f$_{7/2}$ and Pt^{0}4f$_{5/2}$ of the metallic Pt. The smaller peaks at the higher binding energies of Pt^{2+}4f$_{7/2}$ at 72.0 eV, Pt^{2+}4f$_{5/2}$ at 75.3 eV, Pt^{4+}4f$_{7/2}$ at 73.6 eV, and Pt^{4+}4f$_{5/2}$ at 76.9 eV, correspond to PtO and PtO$_2$, respectively [19]. Figure 2c shows the high-resolution Mo 3d spectra, where the two main peaks of Mo 3d spectra are at 229.2 and 232.3 eV, which correspond to the Mo^{4+}3d$_{5/2}$ and Mo^{4+}3d$_{3/2}$ orbitals, and the revealed Mo^{4+} states indicated that the major formation was MoS$_2$ [20]. The other peaks of Mo^{5+}3d$_{5/2}$ at 230.3 eV, Mo^{5+}3d$_{3/2}$ at 233.4 eV, Mo^{6+}3d$_{5/2}$ at 231.4 eV, Mo^{6+}3d$_{3/2}$ at 234.5 eV, Mo^{6+}3d$_{5/2}$ at 232.5 eV, and Mo^{6+}3d$_{3/2}$ at 235.6 eV, corresponded to the minority products of Mo$_2$S$_5$, MoS$_3$, and MoO$_3$, respectively [20,21]. The relatively weak peaks of MoO$_3$ may come from the oxidation of Mo atoms at the edges, or defects on the crystal planes of the MoS$_2$ NSs during the chemical reaction [22]. Figure 2d shows the high-resolution S 2p spectra, the S^{2-}2p$_{3/2}$ and S^{2-}2p$_{1/2}$ peaks at 161.9 and 163.1 eV, which corresponds to MoS$_2$. The S$_2$$^{2-}$2p$_{3/2}$ and S$_2$$^{2-}$2p$_{1/2}$ of binding energy at 163.2 eV and 164.4 eV might represent the intermediate products of Mo$_2$S$_5$ and the MoS$_3$ with a formula of [Mo^{4+}(S$_2$)$^{2-}$S^{2-}] [23,24].

Pt is predominantly present as metallic Pt along with surface oxides and hydroxide, as is normally observed in the case of Pt NPs [19]. Therefore, as above-mentioned, the oxidation groups on the crystal planes of the MoS$_2$ NSs probably act as nucleation sites to reduce the precursor Pt^{4+} to Pt^{2+} and Pt^0 in the mixture solution (H$_2$PtCl$_6$-EG-water) for the subsequent formation of PtNPs by EG reduction.

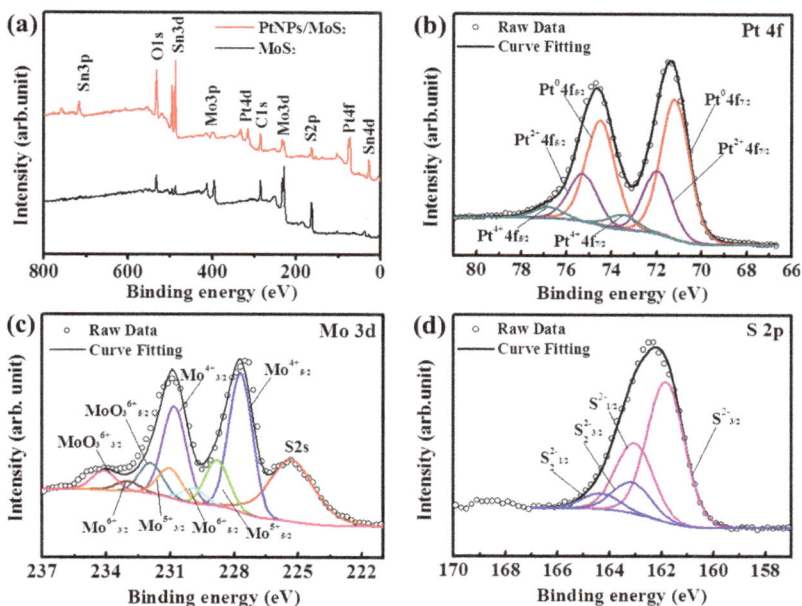

Figure 2. (a) XPS survey spectra; and high-resolution XPS analysis of (b) Pt 4f, (c) Mo 3d, and (d) S 2p of PtNPs/MoS$_2$ NSs hybrids.

Table 1. Mo 3d, S 2p, and Pt 4f peaks in the XPS spectra of PtNPs/MoS$_2$ NSs.

Peak	Fitting of the Peak Binding Energy (eV) and Product					
Pt 4f	Pt^{0}4f$_{7/2}$	Pt^{0}4f$_{5/2}$	Pt^{2+}4f$_{7/2}$	Pt^{2+}4f$_{5/2}$	Pt^{4+}4f$_{7/2}$	Pt^{4+}4f$_{5/2}$
	71.2 (Pt)	74.5 (Pt)	72.0 (PtO)	75.3 (PtO)	73.6 (PtO$_2$)	76.9 (PtO$_2$)
Mo 3d	Mo^{4+}3d$_{5/2}$	Mo^{4+}3d$_{3/2}$	Mo^{5+}3d$_{5/2}$	Mo^{5+}3d$_{3/2}$	Mo^{6+}3d$_{5/2}$	Mo^{6+}3d$_{3/2}$	Mo^{6+}3d$_{5/2}$	Mo^{6+}3d$_{3/2}$
	229.2	232.3	230.3	233.4	231.4	234.5	232.5	235.6
	(MoS$_2$)	(MoS$_2$)	(Mo$_2$S$_5$)	(Mo$_2$S$_5$)	(MoS$_3$)	(MoS$_3$)	(MoO$_3$)	(MoO$_3$)
S 2p	S^{2-}2p$_{3/2}$		S^{2-}2p$_{1/2}$		S^{2-}2p$_{3/2}$		S^{2-}2p$_{1/2}$	
	161.9 (MoS$_2$)		163.1 (MoS$_2$)		163.2 (Mo$_2$S$_5$, MoS$_3$)		164.4 (Mo$_2$S$_5$, MoS$_3$)	

3.2. Electrochemical Properties

EIS measurement was performed in a symmetrical cell comprised of two identical CEs to analyze the correlation between the electrocatalytic activities of the various CEs. Nyquist plots in Figure 3a display the impedance characteristics based on the PtNPs/MoS$_2$ NSs, MoS$_2$ NSs and TD-Pt CEs, and the corresponding electrochemical parameters obtained from the Nyquist plot were fitted with the Autolab FRA software (v4.9, EcoChemie B.V.) and are summarized in Table 2. Based on the equivalent circuit (the inset of Figure 3a), the R_s value was estimated from the intercept on the real axis in the left region, where R_s corresponded to the series resistance of the electrolyte and electrodes. The left semicircle in the high-frequency region corresponded to the charge-transfer resistance (R_{ct}) and the phase angle element (CPE) at the electrolyte-electrode interface, and the right semicircle in the low-frequency region corresponded to the Nernst diffusion impedance (Z_N) in the electrolyte. It is well known that a smaller R_s represents a higher conductivity and a smaller R_{ct} brings a faster charge-transfer rate from the CE to the electrolyte to enhance the electrocatalytic activities. The R_s value corresponds to the series resistance and includes the sheet resistance of the FTO substrate and the resistance of the contacts. The R_s of PtNPs/MoS$_2$ NSs, MoS$_2$ NSs and TD-Pt CEs were 27.8 Ω·cm^2, 27.9 Ω·cm^2, and 27.8 Ω·cm^2, respectively. Based on the R_s results, all CEs had similar conductivities.

The R_{ct} of PtNPs/MoS$_2$ NSs, MoS$_2$ NSs and TD-Pt CEs were 0.75 Ω·cm^2, 12.15 Ω·cm^2 and 3.81 Ω·cm^2, respectively. The results demonstrated that PtNPs/MoS$_2$ NSs CE had outstanding charge transfer ability, that the R_{ct} value of PtNPs/MoS$_2$ NSs CE was 16 times better than that of MoS$_2$ NSs CE, and five times better than that of TD-Pt CE.

In addition, Figure 3b shows the Bode plots used to investigate the charge-transfer kinetics of the various CEs where the high-frequency peak (f_{ct}) and the low-frequency peak (f_N) corresponded to the charge-transfer behavior of the catalytic material and the Nernst-diffusion behavior of the electrolyte, respectively. A higher charge-transfer frequency indicated a lower electron lifetime and led to a lower R_{ct} value for the charge transfer rate [25]. The charge-transfer frequency of PtNPs/MoS$_2$ NSs CE was 4.5 kHz, which was higher than that of MoS$_2$ NSs (0.8 kHz) and TD-Pt (4.3 kHz). The highest frequency of the PtNPs/MoS$_2$ NSs CE obtained from the Bode plot coincided with the lowest R_{ct} value obtained from the Nyquist plot (Figure 3a).

Figure 3c shows the cyclic voltammetry (CV) curves of various CEs, which was measured using a three-electrode system with the Pt sheet as the CE, Ag/AgNO$_3$ as the reference electrode, and various CEs as the working electrode. The relative peaks in Figure 3c corresponded to the redox reactions of I$_3^-$/I$^-$ couples (reduction peak current density (I_{red}): I$_3^-$ + 2e$^-$ → 3I$^-$, oxidation peak current density (I_{ox}): 3I$^-$ → I$_3^-$ + 2e$^-$). The electrocatalytic activity and the redox barrier of I$_3^-$/I$^-$ couples was evaluated in terms of its reduction peak current density (I_{red}) and the voltage separation (E_{pp}) of the I_{red} peak to the I_{ox} peak, respectively. The magnitude of I_{red} corresponded to the catalytic activity of a CE for an I$_3^-$ reduction in a DSSC [26], and the value of E_{pp} was negatively correlated with the standard electrochemical rate constant of a redox reaction [27]. As we can see from Figure 3c and Table 2, the I_{red} value of the PtNPs/MoS$_2$ NSs (−2.17 mA·cm^{-2}) was higher than those of the MoS$_2$ NSs (−1.66 mA·cm^{-2}) and TD-Pt (−1.87 mA·cm^{-2}). The E_{pp} of the PtNPs/MoS$_2$ NSs CE showed a relatively lower value of 0.23 V when compared to that of MoS$_2$ NSs (0.27 V) and TD-Pt (0.24 V) CEs. Therefore, the higher I_{red} and lower E_{pp} values meant better electrocatalytic activity of PtNPs/MoS$_2$ NSs CE in DSSC. The results obtained from CV indicated that PtNPs/MoS$_2$ NSs CE had a better charge transport rate when compared with MoS$_2$ NSs and TD-Pt CEs, which was consistent with the EIS results and Bode plots (Figure 3a,b).

Tafel-polarization curves were used to further investigate the catalytic activities of the exchange current densities (J_0) and the limiting current densities (J_{lim}) at the electrolyte-catalyst interface of various CEs. As shown in Figure 3d, the J_0 was approximately calculated by the Tafel linear extrapolation method, which was closely associated with the R_{ct} value (J_0 is inversely proportional to R_{ct}) and the J_{lim} is dependent on the intersection of the cathodic branch and the vertical axis (Equations (1) and (2)) [13].

$$J_0 = \frac{RT}{nFR_{CT}} \tag{1}$$

$$D = \frac{l}{2nFC} J_{lim} \tag{2}$$

where R is the gas constant; T is the temperature; F is the Faraday constant; l is the spacer thickness; C is the concentration of I$_3^-$ species; and n represents the number of electrons transferred in the reduction reaction.

As shown in Figure 3d and Table 2, the PtNPs/MoS$_2$ NSs CE showed the highest values for J_0 (5.2 mA·cm^{-2}) and J_{lim} (12.1 mA·cm^{-2}), which were higher than those of MoS$_2$ NSs CE (J_0 = 2.7 mA·cm^{-2}, J_{lim} = 10.2 mA·cm^{-2}) and TD-Pt CE (J_0 = 4.4 mA·cm^{-2}, J_{lim} = 11.9 mA·cm^{-2}).

According to the results obtained from the Tafel-polarization curves, the highest J_0 and J_{lim} values of the PtNPs/MoS$_2$ NSs CE indicated the lowest R_{ct} value at the electrolyte–electrode interface, which was consistent with the EIS measurements. Furthermore, the highest J_0 and J_{lim} values of the PtNPs/MoS$_2$ NSs CE also coincided with the highest i_{pc} and lowest E_{pp} values obtained from the CV curves, which strongly agreed with the promotion of the I$_3^-$ reduction rate, thus enhancing the catalytic activity.

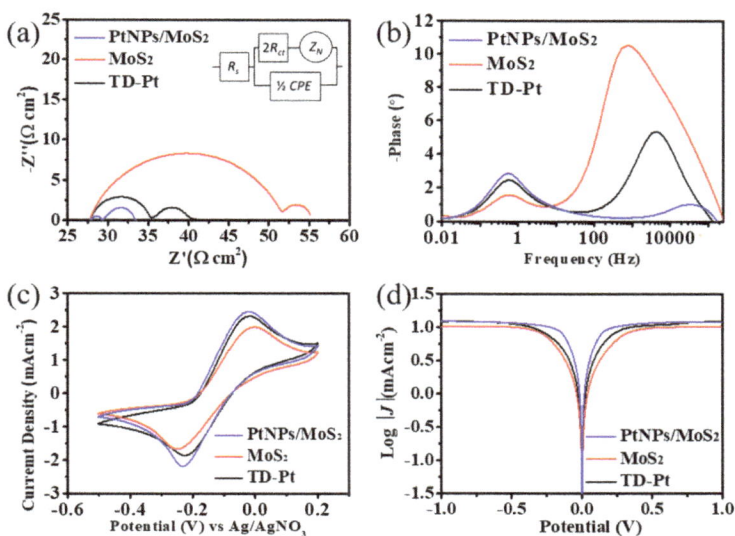

Figure 3. Electrocatalytic properties of various CEs. (**a**) Nyquist plots; (**b**) Bode plots; (**c**) CV curves; and (**d**) Tafel curves.

Table 2. Electrocatalytic properties obtained from Figure 3.

CE	R_s ($\Omega \cdot cm^2$)	R_{ct} ($\Omega \cdot cm^2$)	Z_N ($\Omega \cdot cm^2$)	f_{ct} (kHz)	f_N (Hz)	I_{ox} ($mA \cdot cm^{-2}$)	I_{red} ($mA \cdot cm^{-2}$)	E_{pp} (V)	J_0 ($mA \cdot cm^{-2}$)	J_{lim} ($mA \cdot cm^{-2}$)
PtNPs/MoS$_2$ NSs	27.8	0.75	4.03	4.5	0.57	2.44	−2.17	0.23	5.2	12.1
MoS$_2$ NSs	27.9	12.15	4.24	0.8	0.55	1.98	−1.66	0.27	2.7	10.2
TD-Pt	27.8	3.81	3.93	4.3	0.57	2.31	−1.87	0.24	4.4	11.9

3.3. Photovoltaic Performance of DSSCs

Figure 4 shows the photocurrent-voltage curves of DSSCs assembled with PtNPs/MoS$_2$ NSs, MoS$_2$ NSs, and TD-Pt CEs. The short-circuit current density (J_{sc}), open-circuit voltage (V_{oc}), fill factor (FF), and PCE (η) used to characterize the photovoltaic performances of the DSSCs are summarized in Table 3. The highest J_{sc}, V_{oc}, and FF values of the DSSC for the PtNPs/MoS$_2$ NSs CE were 17.23 mA·cm^{-2}, 0.71 V, and 0.61, respectively, yielding the highest η value of 7.52% out of all the CEs in this work. The J_{sc}, V_{oc}, and FF of the DSSC with the reference TD-Pt CE were 17.13 mA·cm^{-2}, 0.7 V, and 0.57, respectively, yielding an η of 6.92%. The DSSC with the MoS$_2$ NSs CE exhibited an η of 6.76%, and the corresponding J_{sc}, V_{oc}, and FF were 16.12 mA·cm^{-2}, 0.7 V, and 0.59, respectively. It is worth noting that the DSSC based on the MoS$_2$ NSs CE displayed a comparable performance with that obtained using the conventional TD-Pt CE, which can be considered indicative of a surface exposed nanosheet possessing rich catalytic active sites [13].

Table 3. Photovoltaic parameters obtained from Figure 4.

CE	J_{sc} (mA·cm^{-2})	V_{oc} (V)	FF	η (%)
PtNPs/MoS$_2$ NSs	17.23	0.71	0.61	7.52
MoS$_2$ NSs	16.12	0.7	0.59	6.76
TD-Pt	17.13	0.7	0.57	6.92

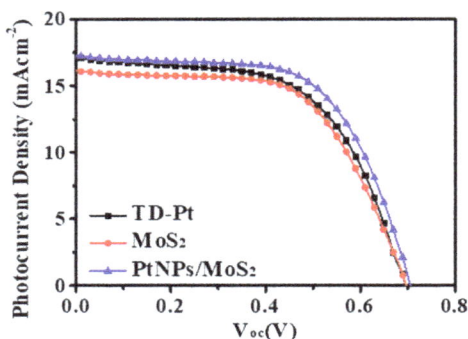

Figure 4. Photocurrent-voltage curves of DSSCs consisted with various CEs.

4. Conclusions

In summary, the PtNPs/MoS$_2$ NSs hybrids exhibited excellent catalytic activities, acting as an outstanding CE for the DSSCs. The two-dimensional nanostructural MoS$_2$ NSs played an important role in this work, as MoS$_2$ NSs not only provided the rich catalytic active sites for the redox reactions, but also displayed a high specific surface area for supporting plentiful Pt NPs. The combination of the two outstanding catalysts showed superior catalytic behavior and led to the superb redox reaction rate for the I$_3^-$/I$^-$ couples. The EIS, Bode plots, CV, and Tafel results all explained the good photovoltaic performance of the DSSCs based on the PtNPs/MoS$_2$ NSs CE, where the lowest R_{ct} value reduced loss during charge transportation, and the highest redox frequency enhanced the charge transfer efficiency, thereby displaying the highest i_{pc}, J_0, and J_{lim} for promoting the charge collection. The superior catalytic properties described above yielded an excellent PCE of 7.52% under AM 1.5 illumination of 100 mW·cm^{-2}, which was 8.7% higher than that of the conventional TD-Pt CE (6.92%).

Acknowledgments: The financial support provided by the Ministry of Science and Technology of Taiwan through Project: MOST 105-2221-E-131-009 is greatly appreciated.

Author Contributions: C.-K.C. designed the methods and experiments, carried out the laboratory experiments, analyzed the data, interpreted the results and wrote the paper. K.-C.H. supported the experiments. T.-K.Y. and C.-K.H. co-worked on developing the conceptual framework and supervised the work. J.-Y.L. supported the electrochemical analysis and discussion. All authors contributed to the revision of the manuscript and have read and approved the final manuscript.

Conflicts of Interest: The authors declare no conflict of interest.

References

1. Oregan, B.; Gratzel, M. A low-cost, high-efficiency solar-cell based on dye-sensitized colloidal TiO$_2$ films. *Nature* **1991**, *353*, 737–740. [CrossRef]
2. Galliano, S.; Bella, F.; Gerbaldi, C.; Falco, M.; Viscardi, G.; Gratzel, M.; Barolo, C. Photoanode/electrolyte interface stability in aqueous dye-sensitized solar cells. *Energy Technol.* **2017**, *5*, 300–311. [CrossRef]
3. Dissanayake, M.A.K.L.; Kumari, J.M.K.W.; Senadeera, G.K.R.; Thotawatthage, C.A.; Mellander, B.E.; Albinsson, I. A novel multilayered photoelectrode with nitrogen doped TiO$_2$ for efficiency enhancement in dye sensitized solar cells. *J. Photochem. Photobiol. A Chem.* **2017**, *349*, 63–72. [CrossRef]
4. Bella, F.; Pugliese, D.; Zolin, L.; Gerbaldi, C. Paper-based quasi-solid dye-sensitized solar cells. *Electrochim. Acta* **2017**, *237*, 87–93. [CrossRef]
5. Bella, F.; Galliano, S.; Falco, M.; Viscardi, G.; Barolo, C.; Gratzel, M.; Gerbaldi, C. Approaching truly sustainable solar cells by the use of water and cellulose derivatives. *Green Chem.* **2017**, *19*, 1043–1051. [CrossRef]

6. Shanti, R.; Bella, F.; Salim, Y.S.; Chee, S.Y.; Ramesh, S.; Ramesh, K. Poly(methyl methacrylate-*co*-butyl acrylate-*co*-acrylic acid): Physico-chemical characterization and targeted dye sensitized solar cell application. *Mater. Des.* **2016**, *108*, 560–569. [CrossRef]

7. Bella, F.; Galliano, S.; Gerbaldi, C.; Viscardi, G. Cobalt-based electrolytes for dye-sensitized solar cells: Recent advances towards stable devices. *Energies* **2016**, *9*, 384. [CrossRef]

8. Chang, L.H.; Hsieh, C.K.; Hsiao, M.C.; Chiang, J.C.; Liu, P.I.; Ho, K.K.; Ma, C.C.M.; Yen, M.Y.; Tsai, M.C.; Tsai, C.H. A graphene-multi-walled carbon nanotube hybrid supported on oxide as a counter electrode of dye-sensitized solar cells. *J. Power Sources* **2013**, *222*, 518–525. [CrossRef]

9. Hu, L.H.; Wu, F.Y.; Lin, C.T.; Khlobystov, A.N.; Li, L.J. Graphene-modified LiFePO$_4$ cathode for lithium ion battery beyond theoretical capacity. *Nat. Commun.* **2013**, *4*. [CrossRef]

10. Radisavljevic, B.; Radenovic, A.; Brivio, J.; Giacometti, V.; Kis, A. Single-layer MoS$_2$ transistors. *Nat. Nanotechnol.* **2011**, *6*, 147–150. [CrossRef] [PubMed]

11. Jaramillo, T.F.; Jorgensen, K.P.; Bonde, J.; Nielsen, J.H.; Horch, S.; Chorkendorff, I. Identification of active edge sites for electrochemical H$_2$ evolution from MoS$_2$ nanocatalysts. *Science* **2007**, *317*, 100–102. [CrossRef] [PubMed]

12. Sun, P.L.; Zhang, W.X.; Hu, X.L.; Yuan, L.X.; Huang, Y.H. Synthesis of hierarchical MoS$_2$ and its electrochemical performance as an anode material for lithium-ion batteries. *J. Mater. Chem. A* **2014**, *2*, 3498–3504. [CrossRef]

13. Al-Mamun, M.; Zhang, H.M.; Liu, P.R.; Wang, Y.; Cao, J.; Zhao, H.J. Directly hydrothermal growth of ultrathin MoS$_2$ nanostructured films as high performance counter electrodes for dye-sensitised solar cells. *RSC Adv.* **2014**, *4*, 21277–21283. [CrossRef]

14. Zhu, Y.Y.; Cui, H.J.; Jia, S.P.; Zheng, J.F.; Wang, Z.J.; Zhu, Z.P. Dynamics investigation of graphene frameworks-supported Pt nanoparticles as effective counter electrodes for dye-sensitized solar cells. *Electrochim. Acta* **2015**, *178*, 658–664. [CrossRef]

15. Yeh, M.H.; Lin, L.Y.; Su, J.S.; Leu, Y.A.; Vittal, R.; Sun, C.L.; Ho, K.C. Nanocomposite graphene/Pt electrocatalyst as economical counter electrode for dye-sensitized solar cells. *Chemelectrochem* **2014**, *1*, 416–425. [CrossRef]

16. Wan, T.H.; Chiu, Y.F.; Chen, C.W.; Hsu, C.C.; Cheng, I.C.; Chen, J.Z. Atmospheric-pressure plasma jet processed Pt-decorated reduced graphene oxides for counter-electrodes of dye-sensitized solar cells. *Coatings* **2016**, *6*, 44. [CrossRef]

17. Yen, M.Y.; Teng, C.C.; Hsiao, M.C.; Liu, P.I.; Chuang, W.P.; Ma, C.C.M.; Hsieh, C.K.; Tsai, M.C.; Tsai, C.H. Platinum nanoparticles/graphene composite catalyst as a novel composite counter electrode for high performance dye-sensitized solar cells. *J. Mater. Chem.* **2011**, *21*, 12880–12888. [CrossRef]

18. Lim, J.; Kim, H.A.; Kim, B.H.; Han, C.H.; Jun, Y. Reversely fabricated dye-sensitized solar cells. *RSC Adv.* **2014**, *4*, 243–247. [CrossRef]

19. Anumol, E.A.; Kundu, P.; Deshpande, P.A.; Madras, G.; Ravishankar, N. New insights into selective heterogeneous nucleation of metal nanoparticles on oxides by microwave-assisted reduction: Rapid synthesis of high-activity supported catalysts. *ACS Nano* **2011**, *5*, 8049–8061. [CrossRef] [PubMed]

20. Kibsgaard, J.; Chen, Z.B.; Reinecke, B.N.; Jaramillo, T.F. Engineering the surface structure of MoS$_2$ to preferentially expose active edge sites for electrocatalysis. *Nat. Mater.* **2012**, *11*, 963–969. [CrossRef] [PubMed]

21. Liu, C.J.; Tai, S.Y.; Chou, S.W.; Yu, Y.C.; Chang, K.D.; Wang, S.; Chien, F.S.S.; Lin, J.Y.; Lin, T.W. Facile synthesis of MoS$_2$/graphene nanocomposite with high catalytic activity toward triiodide reduction in dye-sensitized solar cells. *J. Mater. Chem.* **2012**, *22*, 21057–21064. [CrossRef]

22. Yuwen, L.H.; Xu, F.; Xue, B.; Luo, Z.M.; Zhang, Q.; Bao, B.Q.; Su, S.; Weng, L.X.; Huang, W.; Wang, L.H. General synthesis of noble metal (Au, Ag, Pd, Pt) nanocrystal modified MoS$_2$ nanosheets and the enhanced catalytic activity of Pd-MoS$_2$ for methanol oxidation. *Nanoscale* **2014**, *6*, 5762–5769. [CrossRef] [PubMed]

23. Wang, H.W.; Skeldon, P.; Thompson, G.E.; Wood, G.C. Synthesis and characterization of molybdenum disulphide formed from ammonium tetrathiomolybdate. *J. Mater. Sci.* **1997**, *32*, 497–502. [CrossRef]

24. Weber, T.; Muijsers, J.C.; Niemantsverdriet, J.W. Structure of amorphous MoS$_3$. *J. Phys. Chem.* **1995**, *99*, 9194–9200. [CrossRef]

25. Choi, H.; Kim, H.; Hwang, S.; Han, Y.; Jeon, M. Graphene counter electrodes for dye-sensitized solar cells prepared by electrophoretic deposition. *J. Mater. Chem.* **2011**, *21*, 7548–7551. [CrossRef]

26. Hsieh, C.K.; Tsai, M.C.; Su, C.Y.; Wei, S.Y.; Yen, M.Y.; Ma, C.C.M.; Chen, F.R.; Tsai, C.H. A hybrid nanostructure of platinum-nanoparticles/graphitic-nanofibers as a three-dimensional counter electrode in dye-sensitized solar cells. *Chem. Commun.* **2011**, *47*, 11528–11530. [CrossRef] [PubMed]
27. Roy-Mayhew, J.D.; Bozym, D.J.; Punckt, C.; Aksay, I.A. Functionalized graphene as a catalytic counter electrode in dye-sensitized solar cells. *ACS Nano* **2010**, *4*, 6203–6211. [CrossRef] [PubMed]

coatings

MDPI

Article

Photovoltaic Effect in Graphene/MoS$_2$/Si Van der Waals Heterostructures

Weilin Shi and Xiying Ma *

Suzhou University of Science and Technology, Suzhou 215011, China; goldlionwl@163.com
* Correspondence: maxy@mail.usts.edu.cn; Tel.: +86-512-6841-2662

Academic Editors: Federico Cesano and Domenica Scarano
Received: 16 October 2017; Accepted: 19 December 2017; Published: 21 December 2017

Abstract: This paper presents a study on the photovoltaic effect of a graphene/MoS$_2$/Si double heterostructure, grown by rapid chemical vapor deposition. It was found that the double junctions of the graphene/MoS$_2$ Schottky junction and the MoS$_2$/Si heterostructure played important roles in enhancing the device's performance. They allowed more electron-hole pairs to be efficiently generated, separated, and collected in the graphene/MoS$_2$/Si double interface. The device demonstrated an open circuit voltage of 0.51 V and an energy conversion efficiency of 2.58% under an optical illumination of 500 mW/cm^2. The photovoltaic effect of the device was partly attributed to the strong light absorption and photoresponse of the few-layer MoS$_2$ film, and partly ascribed to the high carrier-collection-rate of the double van der Waals heterostructures (vdWHs) in the device.

Keywords: graphene/MoS$_2$/Si heterostructure; chemical vapor deposition; energy conversion efficiency; photoresponse

---

1. Introduction

Recently, van der Waals heterostructures (vdWHs), consisting of various two-dimensional (2D) materials such as graphene, transition metal sulfide (MoS$_2$, WSe$_2$, etc.) and boron nitride (h-BN), have given rise to many interesting results and phenomena [1–3]. Not only have horizontal and vertical vdWHs increased our physical understanding of 2D systems, but can also be used to fabricate various novel 2D devices, such as vertical transistors, field effect transistors, photodetectors, and photoresponsive memory devices [4–6]. In particular, the graphene/MoS$_2$ heterostructure has emerged as a potential candidate for various novel optoelectronic devices because of the excellent optical properties of MoS$_2$, the high transparency of graphene, and the tunability of its Fermi level [7–9]. Despite the many advantages of the 2D-vdWHs, it was found that the photoelectric conversion efficiency of such optoelectronic devices is no higher than that of the bulk devices. For example, Furchi et al. and Gong et al. reported a conversion efficiency of approximately 0.2% [10] and 0.1% [11] with the MoS$_2$/WSe$_2$ and WSe$_2$/MoSe$_2$ heterostructures, respectively. 2D devices also exhibit a pronounced response to optical signals. This is because, although the 2D materials have good photoresponse characteristics, their weak photoelectron collection ability leads to a high recombination rate and low conversion efficiency. Therefore, it is necessary to further enhance the collection rate of the photogenerated carriers, thereby increasing the photoelectric conversion efficiency. In this work, we designed a graphene/MoS$_2$/Si double heterostructure, that is, the graphene/MoS$_2$ Schottky junction and the MoS$_2$/Si heterostructure. These were used to improve the collection rate of photogenerated carriers, as well as the photoelectric conversion efficiency. The graphene/MoS$_2$/Si double vdWHs were prepared by a rapid chemical vapor deposition, and the photovoltaic effect was investigated. A small Schottky barrier was formed between the few-layer graphene and MoS$_2$ layer due to the difference in their work function, and a large pn barrier was formed between MoS$_2$ and Si substrate, which accelerated the separation of the photogenerated electron–hole pairs and enhanced the carrier collection capability, thus providing a better optical and electrical performance.

2. Materials and Experiment

The structure of the studied graphene/MoS_2/p-Si vdWHs was composed of a graphene layer, a MoS_2 layers, a p-Si, and Ni, Al electrodes; the schematic and the energy band diagram is shown in Figure 1a,b. In Figure 1b, E_{cs}, E_{vs}, E_{Fs}, and χ_s are the conduction band, valence band, Fermi level, and electron affinity, respectively, of Si, while E_{cm}, E_{vm}, E_{Fm}, and χ_m are the conduction band, valance band, Feimi level and electron affinity, respectively, of MoS_2. For Si, χ_s = 4.0 eV, E_{gs} = 1.12 eV. For MoS_2, χ_m = 4.05 eV, E_{gs} = 1.4 eV [12]. E_0 is the vacuum level, and qV is the built-in potential between p-Si and MoS_2. The graphene and MoS_2 layers were fabricated on a p-Si surface by a rapid chemical vapor deposition (CVD) process. The growth system was made up of a large horizontal quartz tube furnace, a vacuum system, a gas meter, and an automatic temperature controller. p-Si (100) substrates with a size of 12 mm × 12 mm × 500 mm were cleaned ultrasonically with a sequence of acetone, ethanol, and deionized water, blown with N_2 to dry them, and finally placed at the center of the furnace. Prior to deposition, the furnace was pumped to 10^{-2} Pa, and heated to 300 °C for 10 min to remove any water moisture. 1 g Analytical grade MoS_2 micro powder and 1 g silver nitrate ($AgNO_3$) powder were dissolved in 200 mL of 5% diluted sulfuric acid (H_2SO_4) by stirring for 5 min at 70 °C in a water bath. Here, H_2SO_4 is a solvent to dissolve MoS_2 powder to become saturated solution, and $AgNO_3$ is a dopant to dope MoS_2 film in situ, which can effectively improve the electrical conductivity of MoS_2 film [13]. Argon (Ar) gas was passed through the mixed solution at a flow rate of 10 cm^3/min to carry $AgNO_3$ and MoS_2 into the reaction furnace. Growth was carried out for 5–10 min, followed by in situ annealing at 800 °C for 30 min. Then, a mixture of pure CH_4 gas (99.999%) and Ar gas with a volume ratio of 1:10 was introduced into the reactive chamber, where the temperature was kept at 950 °C. CH_4 was initially decomposed to give a mixture of C and H_2, and the C atoms condensed to form graphene film [14]. The growth process was carried out for 5 min, and then the samples were annealed at 1000 °C for 30 min. Finally, the samples were removed out when the system had cooled down to room temperature, Ni electrodes (2 mm × 2 mm, 300 nm) were formed by a sputtering method through a shadow mask at the corner of the graphene film, and Al electrode were deposited on the backside of the Si substrate.

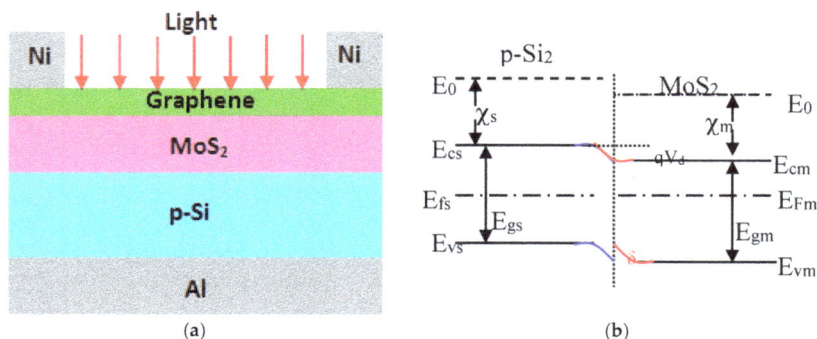

Figure 1. (**a**) The schematics of the grapheme/MoS_2/Si heterostructure device; (**b**) the energy band diagrams of the MoS_2/p-Si vdWHs.

The morphology and structure of the samples were characterized by atomic force microscopy image (AFM, MultiMode 8, Bruker, Billerica, MA, USA) and Raman spectroscopy (Spex-1403, INC, Metuchen, NJ, USA). The optical transparency profiles of the graphene and MoS_2 films were investigated by UV-Vis spectroscopy (UV3600, Shimadzu, Kyoto, Japan). Finally, the photovoltaic characteristics of the graphene/MoS$_2$/p-Si structure were evaluated by a Keithley (Cleveland, OH, USA) 4200 SCS under white light illumination.

3. Result and Discussion

Figure 2a shows an atomic force microscopy (AFM) image of the deposited graphene film. A large uniform graphene film was formed on the substrate, on which some newly generated slices were scattered. From the scale picture in Figure 2b, we can estimate that the thickness of the yellow uniform graphene is approximately 2.5 nm, equaling a few layers of graphene, the highest height of the film is about 5.66 nm. Figure 1c shows the Raman spectrum of the deposited graphene film. We can see that two major scattering peaks appear in the spectrum, a 2D-band peak at 2692 cm^{-1} and a G-band peak at 1580 cm^{-1}. The intensity ratio of $I_G:I_{2D} = 2$ confirms that this is a few-layer graphene [15].

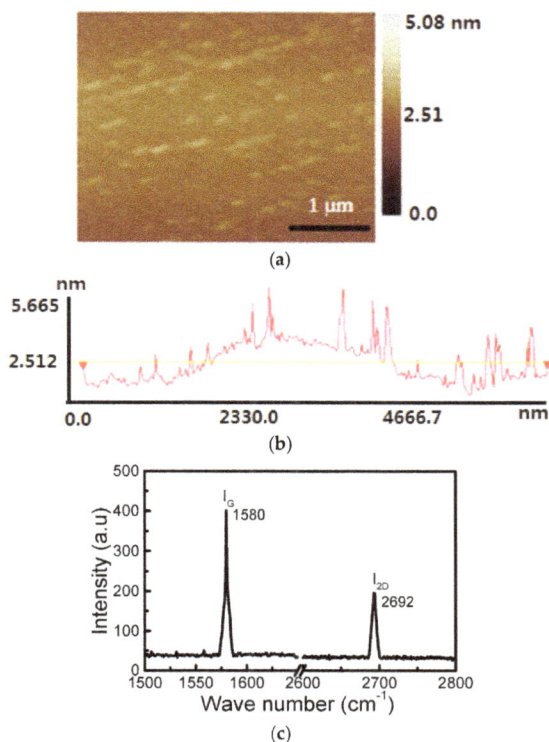

Figure 2. (a) Shows an atomic force microscopy (AFM) image of the deposited graphene film; (b) the scale bar of the graphene; (c) the Raman spectrum of the deposited graphene.

Figure 3a shows the AFM image of MoS$_2$ film deposited on the Si substrate. It is seen that many MoS$_2$ rings with thicknesses of approximately 5 nm are uniformly distributed on the surface, which is equivalent to ten layers of MoS$_2$. The sublayer MoS$_2$ is a homogeneous, continuous film with a thickness of approximately 5 nm. The rings take a circular shape with an inner and outer diameter of 50 and 100 nm respectively. They are uniformly scattered on the substrate without any overlap, dislocation or defect, indicating that they are formed by an excellent self-organized growth process with a layer-by-layer growth mode. Figure 2b shows the Raman spectrum of MoS$_2$ with two typical strong waving peaks at 385 cm^{-1} and 406 cm^{-1}, corresponding to the in-plane (E_{2g}^1) and the out-of-plane (A_{1g}) modes respectively [16]. It has been reported that the E_{2g}^1 mode is attenuated and the A_{1g} mode is strengthened with increasing layer thickness [16], which is similar to other layered materials, where the bond distance changes with number of layers [17]. The frequency difference of the Raman modes of A_{1g} and E_{2g}^1 is about 21 cm^{-1}, indicating that the deposited MoS$_2$ is few-layered, based on [18,19].

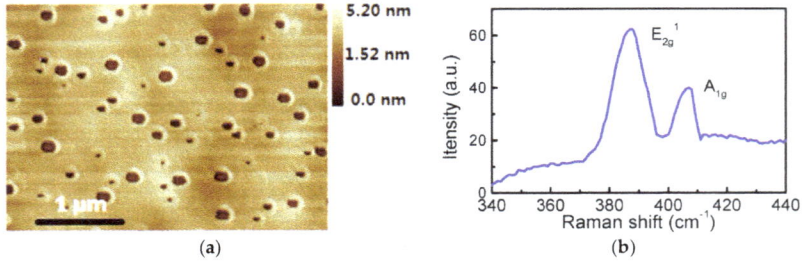

Figure 3. (**a**) The AFM image of MoS$_2$ film deposited on Si substrate; (**b**) the Raman spectrum of MoS$_2$.

The transmittance of the graphene sample in the visible light range of 400–800 nm is shown in Figure 4a. The optical transparency of the graphene deposited for 5 min was over 80% in the visible range. Moreover, the transparency increases with wavelength, becoming almost fully transparent for the range of 600–800 nm. A high transmittance is very useful for making solar cells, because light in the 400–800 nm range has higher power. Theoretically, the transparency of graphene drops quickly with thickness [20]. However, the actual measured transparency of graphene does not closely obey this pattern. Wang et al. reported that the transparency of GO is over 80% in 550 nm light at a thickness of 22 to 78 nm [21]. Figure 3b shows the UV-Vis absorption spectra of the MoS$_2$/Si heterojunction sample in the wavelength range of 250–900 nm. The absorption peak for the MoS$_2$ film, corresponding to the band gap of MoS$_2$ (approximately 1.69 eV), is located at 735 nm, and the absorption peak of Si occurs at 850 nm. It is clear that the optical absorption of the MoS$_2$/Si heterojunction covered the visible and near-infrared spectral regions of 350–900 nm, which could help to improve the efficiency of solar cells.

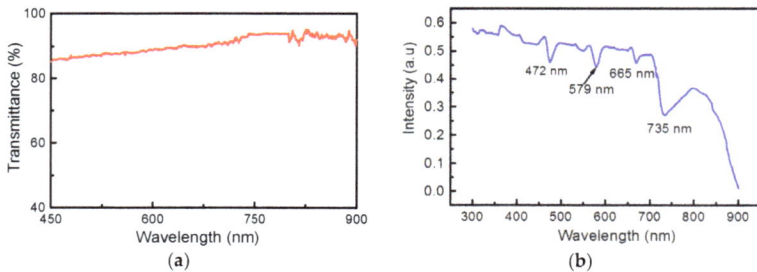

Figure 4. (**a**) The transmittance the visible light range of the graphene sample; (**b**) the UV-Vis absorption spectra of the MoS$_2$/Si heterojunction sample in the wavelength range 250–900 nm.

Figure 5a,b shows the current-voltage (I–V) characteristic curves of the graphene/MoS$_2$/Si heterojunction solar cell under different light energy densities without and with illumination, respectively. Without illumination, the positive current I increases exponentially with the applied voltage for $V > 0$, while the reverse current reduces to almost zero when $V < 0$. The heterojunction shows very low reverse saturation current and rectification properties. This shows that the heterojunction has good interface and contact properties, although there is a large lattice mismatch between MoS_2 and graphene. No reconstruction is expected when they contact closely [22], because there are few dangling bonds and surface states in the two dimensional films, and they are compacted by van der Waals interaction with minimum strain. The *I–V* curves measured with optical power densities from 100 to 500 mW/cm² are presented in Figure 5b. The reverse current is much higher here than that during darkness, and all the curves in reverse bias show good saturation characteristics. Clearly, the current increases with increasing light energy. The heterojunction has a large open circuit voltage (*V*oc) of 0.51 V, and a short-circuit current of 0.51 µA at an illumination of 500 mW/cm².

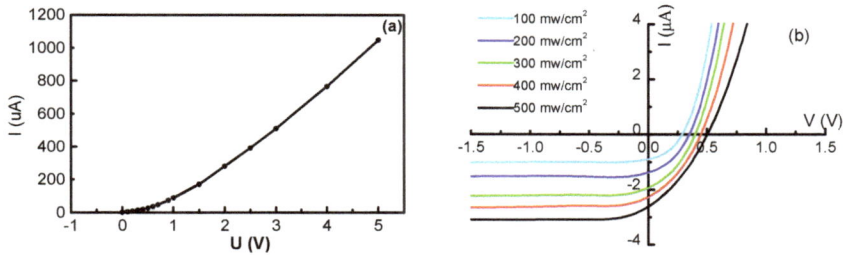

Figure 5. The current-voltage (*I–V*) curve of the graphene/MoS_2/Si heterojunctions in darkness (**a**), and illumination with different light energies (**b**).

A Schottky barrier forms between the graphene and MoS_2 film, and a pn heterostructure forms between the MoS_2 film and Si substrate. Since the graphene layer has high transmission properties, only a small part of the light will be absorbed by the few-layer graphene, while a large part will pass through. The transmitted light is further absorbed by the MoS_2 film and Si substrate, which produces intrinsic absorption when the photon energy is larger than the optical band gap, giving rise to photogenerated electron–hole pairs. The holes, being the minority carrier in the MoS_2 layer, have a concentration gradient, which will diffuse to the space-charge region boundary of MoS_2/Si heterojunction. Thereafter, they are accelerated to the Si side under the built-in electric field in the space-charge region. Thus, the photogenerated electrons and holes are separated by the built-in electric field in the space, and they accumulate on the MoS_2 and Si sides, respectively, generating the photovoltaic effect between the MoS_2 and Si surface.

The short-circuit current (I_{SC}) and the open-circuit voltage (V_{OC}) of the graphene/MoS_2/Si heterojunction solar cell with different light energies are shown in Figure 6a,b, respectively. V_{OC} represents the voltage when no current is flowing through the device, while I_{SC} shows the current at zero voltage between the electrodes. Generally, I_{SC} and V_{OC} linearly increase with light energy because the photogenerated carriers are in direct proportion to the light energy [23]. We can see that the conversion of optical energy into electrical energy of the graphene/MoS_2/Si device is approximately 2.5%. It is larger than the previously reported values of 2.15% and 1.47% for a monolayer graphene film/Si-nanowire-array Schottky junction solar cell [16] and graphene nanoribbon/multiple-silicon-nanowires junctions [24], respectively. Moreover, it is almost ten times larger than that of the MoS_2/WSe_2 [10] and the WSe_2/$MoSe_2$ Heterostructures [11]. The high conversion efficiency in our work may be attributed to the following reason. Firstly, electrons in the graphene layer have high mobility, which imparts good conductivity and low contact resistance. Secondly, the double junctions can accelerate the separation process of the photogenerated electrons and holes, thereby decreasing their recombination. Finally, the bulk Si substrate has high carrier collection capability, which can effectively enhance the conversion efficiency.

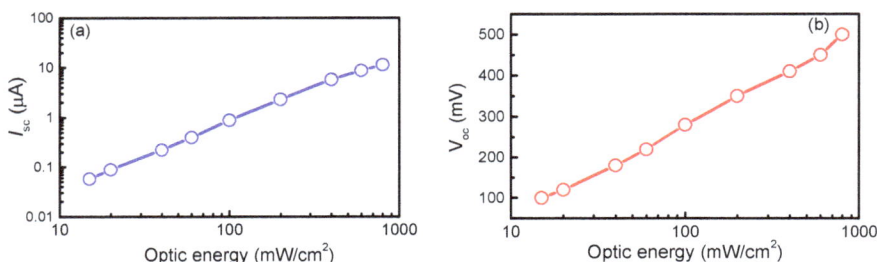

Figure 6. The parameter dependence of the graphene/MoS$_2$/Si heterojunctions with optical power density: (**a**) The short-circuit current (I_{SC}); (**b**) the open-circuit voltage (V_{OC}).

From the results, we deposited a large area of uniform graphene and MoS$_2$ films by chemical vapor deposition. The graphene/MoS$_2$/Si heterojunctions have good *I–V* properties, photovoltaic effect and high photoelectric conversion efficiency, showing that the heterojunction has good contacting and interface properties, which could be used to fabricate high-efficiency solar cells, optical detectors, and other light-response devices.

4. Conclusions

Few-layer graphene and MoS$_2$ films were prepared on a Si substrate by rapid chemical vapor deposition, and the photovoltaic effects of the graphene/MoS$_2$/Si double-junction were investigated. The graphene and MoS$_2$ films have a large-area uniform morphology with a thickness of tens of monolayers. The graphene/MoS$_2$ heterojunction has good interface and photoelectric response properties. An open-circuit voltage of 0.51 V and an energy conversion efficiency of 2.58% were achieved under an illumination energy of 500 mW/cm^2. The higher conversion efficiency was attributed to the double junction and Si substrate, which played important roles in collecting carriers and enhancing the conversion efficiency. The double-junction device can harvest solar light and generate more electron–hole pairs efficiently.

Acknowledgments: This work is supported in part by the National Natural Science Foundation of China (No. 31570515) and the Scientific Project Program of Suzhou City (No. SYN201511).

Author Contributions: Weilin Shi contributed with the experimental components regarding tool preparation, sample deposition and measurement; Xiying Ma designed the experiments, analyzed the data and wrote the paper.

Conflicts of Interest: The authors declare no conflict of interest. The founding supporters had no role in the design of the study; in the collection, analyses, or interpretation of data; in the writing of the manuscript, and in the decision to publish the results.

References

1. Shih, C.; Wang, Q.H.; Son, Y.; Jin, Z.; Blankschtein, D.; Strano, M.S. Tuning on-off current ratio and field-effect mobility in a MoS$_2$-graphene heterostructurevia Schottky barrier modulation. *ACS Nano* **2014**, *8*, 5790–5798. [CrossRef] [PubMed]

2. Kwak, J.Y.; Hwang, J.; Calderon, B.; Alsalman, H.; Munoz, N.; Schutter, B.; Spencer, M.G. Electrical characteristics of multilayer MoS$_2$ FET's with MoS$_2$/graphene heterojunction contacts. *Nano Lett.* **2014**, *14*, 4511–4516. [CrossRef] [PubMed]

3. Baugher, B.W.H.; Churchill, H.O.H.; Yang, Y.; Jarillo-herrero, P. Optoelectronic devices based on electrically tunable p–n diodes in a monolayer dichalcogenide. *Nat. Nanotechnol.* **2014**, *9*, 262–267. [CrossRef] [PubMed]

4. Roy, K.; Padmanabhan, M.; Goswami, S.; Sai, T.P.; Ramalingam, G.; Raghavan, S.; Ghosh, A. Graphene–MoS$_2$ hybrid structures for multifunctional photoresponsive memory devices. *Nat. Nanotechnol.* **2013**, *8*, 826–830. [CrossRef] [PubMed]

5. Yu, W.J.; Li, Z.; Zhou, H.; Chen, Y.; Wang, Y.; Huang, Y.; Duan, X. Vertically stacked multi-heterostructures of layered materials for logic transistors and complementary inverters. *Nat. Mater.* **2013**, *12*, 246–252. [CrossRef] [PubMed]

6. Ross, J.S.; Klement, P.; Jones, A.M.; Ghimire, N.J.; Yan, J.; Mandrus, D.G.; Taniguchi, T.; Watanabe, K.; Kitamura, K.; Yao, W.; et al. Electrically tunable excitonic light-emitting diodes based on monolayer WSe_2 p–n junctions. *Nat. Nanotechnol.* **2014**, *9*, 268–272. [CrossRef] [PubMed]

7. Lin, Y.F.; Li, W.; Li, S.L.; Xu, Y.; Aparecido-Ferreira, A.; Komatsu, K.; Sun, H.; Nakaharai, S.; Tsukagoshi, K. Barrier inhomogeneities at vertically stacked graphene-based heterostructures. *Nanoscale* **2014**, *6*, 795–799. [CrossRef] [PubMed]

8. Sinha, D.; Lee, J.U. Ideal graphene/silicon schottky junction diodes. *Nano Lett.* **2014**, *14*, 4660–4664. [CrossRef] [PubMed]

9. Das, S.; Gulotty, R.; Sumant, A.V.; Roelofs, A. All two-dimensional, flexible, transparent, and thinnest thin film transistor. *Nano Lett.* **2014**, *14*, 2861–2866. [CrossRef] [PubMed]

10. Mueller, T.; Furchi, M.M.; Zechmeister, A.; Schuler, S.; Pospischil, A. Atomically-thin van der Waals heterostructure solar cells. In Proceedings of the 2015 Conference on Lasers and Electro-Optics (CLEO), San Jose, CA, USA, 10–15 May 2015.

11. Gong, Y.; Lei, S.; Ye, G.; Li, B.; He, Y.; Keyshar, K.; Zhang, X.; Wang, Q.; Lou, J.; Liu, Z.; et al. Two-step growth of two-dimensional $WSe_2/MoSe_2$ heterostructures. *Nano Lett.* **2015**, *15*, 6135–6141. [CrossRef] [PubMed]

12. Lee, E.W., II; Lee, C.H.; Paul, P.K.; Ma, L.; McCulloch, W.D.; Krishnamoorthy, S.; Wu, Y.; Arehart, A.R.; Rajan, S. Layer-transferred MoS_2/GaN PN diodes. *Appl. Phys. Lett.* **2015**, *107*, 103505. [CrossRef]

13. Gu, W.; Shen, J.; Ma, X. Fabrication and electrical properties of MoS_2 nanodisc-based back-gated field effect transistors. *Nano Res. Lett.* **2014**, *9*, 100–104. [CrossRef] [PubMed]

14. Ma, X.; Gu, W.; Shen, J.; Tang, Y. Investigation of electronic properties of graphene/Si field-effect transistor. *Nano Res. Lett.* **2012**, *7*, 677–681. [CrossRef] [PubMed]

15. Ni, Z.H.; Wang, H.M.; Kasim, J.; Fan, H.M.; Yu, T.; Wu, Y.H.; Feng, Y.P.; Shen, Z.X. Graphene thickness determination using reflection and contrast spectroscopy. *Nano Lett.* **2007**, *7*, 2758–2763. [CrossRef] [PubMed]

16. Lee, Y.H.; Zhang, X.Q.; Zhang, W.; Chang, M.T.; Lin, C.T.; Chang, K.D.; Yu, Y.C.; Wang, J.T.W.; Chang, C.S.; Li, L.J.; et al. Synthesis of large-area MoS_2 atomic layers with chemical vapor deposition. *Adv. Mater.* **2012**, *24*, 2320–2325. [CrossRef] [PubMed]

17. Arenal, R.; Ferrari, A.C.; Reich, S.; Wirtz, L.; Mevellec, J.Y.; Lefrant, S.; Rubio, A.; Loiseau, A. Raman spectroscopy of single-wall boron nitride nanotubes. *Nano Lett.* **2006**, *6*, 1812–1816. [CrossRef] [PubMed]

18. Cravanzola, S.; Muscuso, L.; Cesano, F.; Agostini, G.; Damin, A.; Scarano, D.; Zecchina, A. MoS_2 nanoparticles decorating titanate-nanotube surfaces: Combined microscopy, spectroscopy, and catalytic studies. *Langmuir* **2015**, *31*, 5469–5478. [CrossRef] [PubMed]

19. Lee, C.; Yan, H.; Brus, L.E.; Heinz, T.F.; Hone, J.; Ryu, S. Anomalous lattice vibrations of single-and few-layer MoS_2. *ACS Nano* **2010**, *4*, 2695–2700. [CrossRef] [PubMed]

20. Nair, R.R.; Blake, P.; Grigorenko, A.N.; Novoselov, K.S.; Booth, T.J.; Stauber, T.; Peres, N.M.R.; Geim, A.K. Fine structure constant defines visual transparency of graphene. *Science* **2008**, *320*, 1308–1315. [CrossRef] [PubMed]

21. Wang, S.J.; Geng, Y.; Zheng, Q.; Kim, J.K. Fabrication of highly conducting and transparent, graphene films. *Carbon* **2010**, *48*, 1815–1823. [CrossRef]

22. Lu, C.P.; Li, G.; Watanabe, K.; Taniguchi, T.; Andrei, E.Y. MoS_2: Choice substrate for accessing and tuning the electronic properties of grapheme. *Phys. Rev. Lett.* **2014**, *113*, 156804. [CrossRef] [PubMed]

23. Xie, C.; Lv, P.; Nie, B.; Jie, J.S.; Zhang, X.W.; Wang, Z.; Jiang, P.; Hu, Z.Z.; Luo, L.B.; Zhu, Z.F.; et al. Quantitative measurement of scattering and extinction spectra of nanoparticles by darkfield microscopy. *Appl. Phys. Lett.* **2011**, *99*, 133113. [CrossRef]

24. Giovannetti, G.; Khomyakov, P.A.; Brocks, G.; Karpan, V.M.; van den Brink, J.; Kelly, P.J. Doping graphene with metal contacts. *Phys. Rev. Lett.* **2008**, *101*, 026803. [CrossRef] [PubMed]

coatings

MDPI

Article

Synthesis, Characterization, and Application of Novel Ni-P-Carbon Nitride Nanocomposites

Eman M. Fayyad [1,†], Aboubakr M. Abdullah [1,*], Mohammad K. Hassan [1], Adel M. Mohamed [2], Chuhong Wang [3], George Jarjoura [3] and Zoheir Farhat [3]

1   Center for Advanced Materials, Qatar University, Doha 2713, Qatar; emfayad@qu.edu.qa (E.M.F.); mohamed.hassan@qu.edu.qa (M.K.H.)
2   Department of Metallurgical and Materials Engineering, Faculty of Petroleum and Mining Engineering, Suez University, Suez 43721, Egypt; adel.mohamed25@yahoo.com
3   Department of Mechanical Engineering, Dalhousie University, Halifax, NS B3J 2X4, Canada; iriswang@dal.ca (C.W.); George.Jarjoura@dal.ca (G.J.); Zoheir.Farhat@Dal.Ca (Z.F.)
*   Correspondence: bakr@qu.edu.qa; Tel.: +974-3307-0591; Fax: +974-4403-3889
†   Permanent address: Physical Chemistry Department, National Research Center, Dokki, Giza 12622, Egypt.

Received: 28 November 2017; Accepted: 15 January 2018; Published: 17 January 2018

Abstract: Dispersion of 2D carbon nitride (C_3N_4) nanosheets into a nickel phosphorous (NiP) matrix was successfully achieved by ultrasonication during the electroless plating of NiP from an acidic bath. The morphology and thickness, elemental analysis, phases, roughness, and wettability for as-plated and heat-treated nanocomposite were determined by scanning electron microscopy, energy-dispersive X-ray spectroscopy, X-ray diffraction, atomic force microscopy, and contact angle measurements, respectively. C_3N_4 showed a homogeneous distribution morphology in the nanocomposite that changed from amorphous in case of the NiP to a mixed crystalline-amorphous structure in the NiP-C_3N_4 nanocomposite. The microhardness and corrosion resistance of the as-plated nanocomposite and the heat-treated nanocomposite coating were significantly enhanced compared to the Ni-P. The nanocomposite showed a superior corrosion protection efficiency of ~95%, as observed from the electrochemical impedance spectroscopy (EIS) measurements. On the other hand, the microhardness of the nanocomposite was significantly increased from 780 to reach 1175 HV_{200} for NiP and NiP-C_3N_4, respectively.

Keywords: electroless NiP alloy; carbon nitride; composites coating; corrosion; microhardness

1. Introduction

Enhancing the corrosion protection for oil and gas pipelines continues to motivate intensive research efforts to find new coatings or modify the existing ones. Electroless deposited NiP coatings, obtained by an autocatalytic process, are characterized by a combination of many unique properties such as good wear and corrosion protection efficiency, uniformity of coating thickness, and a higher hardness [1]. These properties opened the field for NiP to be used in different industries [2]. However, to improve these properties, the incorporation of nanoparticles in the NiP matrix has been tried extensively recently [3–8]. The performance of the co-deposition of hard second-phase nanoparticles such as Al_2O_3, TiN, B_4C, ZrO_2, SiC, TiO_2, CNT, graphite, and diamond [9–17] with the Ni-P matrix was investigated, and has shown an enhancement in terms of microhardness, as well as corrosion and wear resistances. Nevertheless, the Ni-P matrix with the insertion of the carbon nitride (C_3N_4) nano-sheets has not been reported. The prediction of the possible existence of the C_3N_4 compound is credited to Cohen and Liu [18,19]. Due to its high hardness and excellent thermal and chemical stability, C_3N_4 has attracted significant interest. C_3N_4 properties are essentially the same as those of diamond [20], i.e., the mechanical and tribological characteristics, as well as the corrosion resistance of its composite coatings, are expected to be excellent [21]. The goal of the present study is to prepare a new NiP-C$_3$N$_4$ nanocomposite with better corrosion resistance and mechanical properties through the incorporation of 2D C$_3$N$_4$ nanosheets during the electroless deposition of the NiP alloy. In addition, a comparative study between the new nanocomposite and the original C$_3$N$_4$-free alloy will be carried out to show the superior performance of the electroless deposited NiP-C$_3$N$_4$ nanocomposite compared to the NiP alloys.

2. Experimental

2.1. Materials, Solutions, and Preparation

Electroless NiP and NiP-C$_3$N$_4$ nanocomposite coatings were deposited on an API X120 C-steel that was starting to be used recently in the oil and gas industry. The chemical composition (in wt %) of the API X120 steel that purchased from Tianjin Tiangang Guanye Co., Ltd., (Tianjin, China) is shown in Table 1.

Table 1. The chemical composition (in wt %) of the substrate.

C	Si	Mn	Ni	Cr	Mo	Cu	V	Fe
0.129	0.101	0.541	0.017	0.039	0.0013	0.015	0.25	balance

Prior to the electroless deposition, the specimens were grinded with different grits of emery paper up to 2000, then polished with micro-polish alumina suspension (1 and 3 μm) to obtain a mirror finishing surface. After that degreasing the specimens in ultrasonicated acetone bath for 15 min was done, followed by alkaline cleaning for 5 min at 80 °C then electro-alkaline cleaning for 2 min at 70 °C with $I = 2$ A·cm^{-2} and acid etching in 15 wt % H$_2$SO$_4$ solution for 20 s. The used alkaline cleaning solution consists of 50 g·L^{-1} NaOH, 30 g·L^{-1} Na$_2$CO$_3$ and 30 g·L^{-1} Na$_3$PO$_4$ while that used in electro-alkaline cleaning consists of 15 g·L^{-1} NaOH, 25 g·L^{-1} Na$_2$CO$_3$ and 25 g·L^{-1} Na$_3$PO$_4$.

After each of the pretreatment steps, the specimens were washed with deionized water. Pretreated substrate coupons of $20 \times 30 \times 10$ mm^3 were used in the electroless deposition process. All used solutions are analytical-grade reagents from Sigma-Aldrich (St. Louis, MO, USA).

The electroless bath for the NiP coating contained 15 g·L^{-1} NiSO$_4$·6H$_2$O, 30 g·L^{-1} NaH$_2$PO$_2$·H$_2$O, 20 g· L^{-1} lactic acid, 20 g·L^{-1} citric acid, and 0.002 g·L^{-1} thiourea, while the electroless NiP-C$_3$N$_4$ nanocomposite coating bath contained 15 g·L^{-1} NiSO$_4$·6H$_2$O, 30 g·L^{-1} NaH$_2$PO$_2$·H$_2$O, 15 g· L^{-1} ammonium chloride, 30 g·L^{-1} sodium citrate, and 0.002 g·L^{-1} thiourea. NaOH was used to adjust the pH of the NiP and NiP-C$_3$N$_4$ plating baths to 4.5 and 8, respectively. The two baths were maintained at 85 °C. To obtain well-suspended C$_3$N$_4$ nanosheets, which are synthesized and characterized in our previous work [22], in the electroless bath, 0.5 g of C$_3$N$_4$ was added to a 100 mL of the plating solution that includes also 0.02 g·L^{-1} of sodium dodecyl sulfate as the surfactant. Then, the solution was mixed with ultrasonic probe for 2 h. Finally, the mixture was poured into the original plating bath stirred at 300 rpm using a magnetic stirrer. The coupons were immersed in the bath and the plating process lasted for 2 h. Then, the coupons were removed from the bath, rinsed with deionized water, and dried with blowing air. After plating, three samples of each coat underwent heat treatment (HT) at 400 °C for 1 h under vacuum to study the change in the properties of the NiP and NiP-C$_3$N$_4$ nanocomposite after the HT.

2.2. Characterization

The surface morphology and the elemental analysis of the nanocomposite were performed using a scanning electron microscope (SEM, Nova NanoSEM 450, Thermo Fisher Scientific, Eindhoven, The Netherlands) coupled with an energy-dispersive X-ray spectroscopy (EDX, Bruker detector 127 eV, Bruker, Leiderdorp, The Netherlands). The analysis of the different phases of the nanocomposite coating was performed using X-ray diffractometry (XRD, Miniflex2 Desktop, Cu Kα,

Rigaku, Tokyo, Japan). A Vickers microhardness tester (FM-ARS9000, Future-Tech Corp., Tokyo, Japan) was used to measure the microhardness at a load of 200 g for 10 s. The measurements were repeated five times on each sample and the mean of the results was considered. Atomic Force Microscopy (AFM) was used to inspect the heterogeneities (surface roughness) of the coated specimens. An MFP3D Asylum research (Asylum Research, Santa Barbara, CA, USA) AFM equipped with a silicon probe (Al reflex coated Veeco model–OLTESPA, Olympus, Tokyo, Japan; Spring constant: 2 N·m^{-1}, resonant frequency: 70 kHz) was utilized for all roughness experiments. Measurements were conducted at ambient conditions using the Standard Topography AC in air (tapping mode in air). Contact angle measurements were conducted using a DataPhysics OCA35, DataPhysics Instruments GmbH, Filderstadt, Germany). Four microliters of deionized water were used as the probing liquid. To achieve accurate results, at least five contact angles were measured, and their average is reported.

2.3. Corrosion Study

The electrochemical impedance spectroscopy (EIS) was performed using a three-electrode cell at 25 °C in a 3.5% (*w/w*) NaCl solution utilizing a Gamry electrochemical workstation (Gamry Instruments, Warminster, PA, USA). The corrosion resistance of the electroless-plated NiP and the NiP-C$_3$N$_4$ nanocomposite was examined. An area of 2 cm^2 of the coated specimen was the working electrode, while an Ag/AgCl electrode and a graphite rod were used as the reference and the counter electrodes, respectively. A 10 mV AC amplitude was used, and the frequency varied from 0.01 Hz to 100 kHz. All EIS data were recorded after a steady-state, open-circuit potential was achieved. A 3.5 wt % sodium chloride solution was utilized to expose a 2 cm^2 area of the surface. Following the immersion of the specimens in a 3.5% (*w/w*) NaCl solution for 3 h and keeping them at an open-circuit potential for 20 min, potentiodynamic polarization tests (Tafel analysis) were performed. A scan rate of 0.167 mV·s^{-1} and a potential range of ±250 mV vs. the open circuit potential was used to acquire the anodic and cathodic polarization curves.

3. Results and Discussion

3.1. Surface Morphology of the Ni-P and Ni-P-C$_3$N$_4$ Coatings

The SEM surface morphology of the as-plated NiP and NiP-C$_3$N$_4$ nanocomposite layers are represented in Figure 1a,b, respectively. The surfaces of both coatings exhibit a typical "cauliflower-like" morphology that shows many grains, and each contains many fine granules. Although the cauliflower morphology is a common feature for the electroless-plated NiP alloys, it can be observed that there are spaces between the nodules that facilitate the diffusion of the corrosive media to the substrate, as shown in Figure 1a. Figure 1b reveals a fine microstructure of the NiP-C$_3$N$_4$ nanocomposite coating compared to the NiP coatings that are shown in Figure 1a. This may be attributed to the uniform distribution of the C$_3$N$_4$ nanosheets in the NiP composite coating surface that results in a more fine compact structure with higher surface roughness, as shown later in AFM measurements. Figure 1c,d showed the EDX spectra of the as-plated NiP and NiP-C$_3$N$_4$ coatings that illustrate the presence of nitrogen in the nanocomposite, which proved the successful co-deposition of the C$_3$N$_4$ nanosheets into the NiP matrix. As seen in Figure 1e, for more illustration, the EDX mapping proves that the C$_3$N$_4$ nanosheets are uniformly and homogeneously distributed in the NiP matrix.

Furthermore, examination of Figure 1c,d shows that the NiP and NiP-C$_3$N$_4$ coatings contain approximately 10.48 and 8.76 wt % phosphorus, respectively. According to [23,24], the microstructure of ENP (Electroless NiP) coatings has been reported to be either amorphous or crystalline, or both, depending on the phosphorous content involved. For instance, crystalline, mixed amorphous-crystalline, and amorphous structures have been reported for low (1–5 wt %), medium (6–9 wt %), and high (10–13 wt %) phosphorous ENP coatings, respectively. Therefore, the microstructure of the NiP coating is amorphous (see XRD results in Section 3.3), whereas that of the NiP-C$_3$N$_4$ coating tends to be a mixed crystalline-amorphous, as depicted by XRD results in Section 3.3.

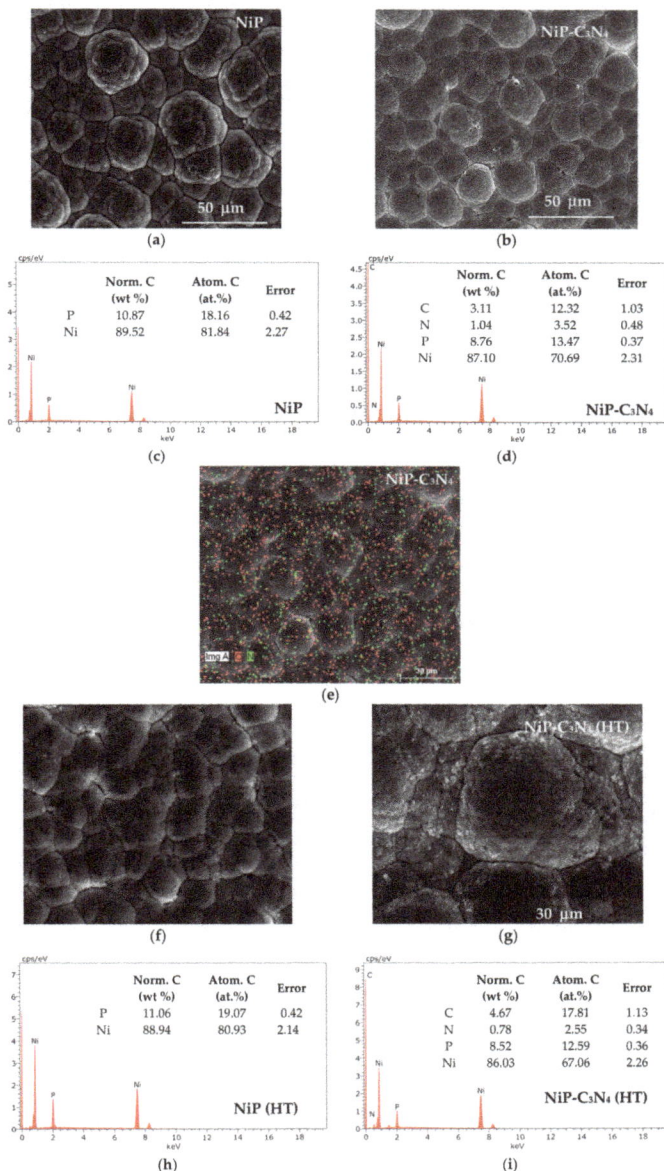

(a)

(b)

(c)

(d)

(e)

(f)

(g)

(h)

(i)

Figure 1. (**a**,**b**,**f**,**g**) SEM micrographs and (**c**,**d**,**h**,**i**) EDX elemental analysis of the electroless as-plated (**a**,**c**) NiP, (**b**,**d**) as-plated NiP-C$_3$N$_4$, (**f**,**h**) heat-treated NiP, and (**g**,**i**) heat-treated NiP-C$_3$N$_4$, respectively; (**e**) EDX mapping for the C and N elements on the surface of the NiP-C$_3$N$_4$ composite coating.

After HT at 400 °C for 1 h, the granular morphology of the as-plated NiP coating is gradually diminished and becomes smoother, as shown in Figure 1f. Comparison of the as-plated NiP-C$_3$N$_4$

composite coating to the heat-treated composite coating reveals that there is a difference in the nodular morphology, and the particles become larger and more agglomerated, as shown in Figure 1g. The change in the morphology with HT may be attributed to the different diffusion coefficients for the different phases in the composite coatings.

The cross-section morphology of as-plated NiP and NiP-C$_3$N$_4$ coatings was checked using SEM, as shown in Figure 2. It was revealed that the thickness of the NiP coating is approximately 52 µm, whereas the NiP-C$_3$N$_4$ coating has a thickness of 35 µm. This indicates that the existence of the reinforcing phase (C$_3$N$_4$) in the coating decreased the thickness considerably. Decreasing the thickness reveals the low deposition rate of the NiP-C$_3$N$_4$ coating compared to that for the NiP one. This may be attributed to the possibility of the physical adsorption of some C$_3$N$_4$ particles on the catalytic surface that result in the minimization of the available active sites for the deposition process that decreases the overall deposition rate [25]. In addition, there are no defects or cracks observed at the substrate-coating interface. This demonstrates the good adhesion of the coatings. The thickness of both coatings does not change after the HT.

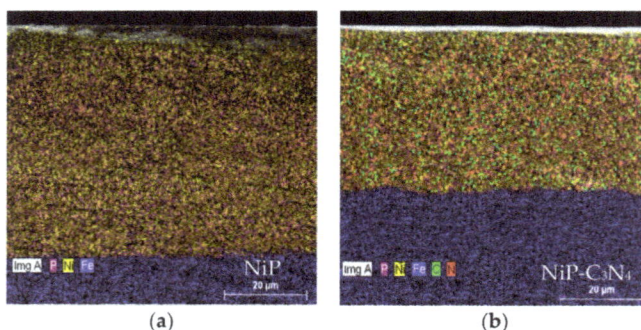

Figure 2. The cross-sectional SEM-photomicrographs of the (**a**) Ni-P and (**b**) NiP-C$_3$N$_4$ nanocomposite coatings.

3.2. Structures of the Ni-P and Ni-P-C$_3$N$_4$ Coatings

XRD patterns of both as-plated and heat-treated Ni-P and Ni-P-C$_3$N$_4$ layers are represented in Figure 3a,b, respectively. It is observed that the diffraction pattern of both NiP and NiP-C$_3$N$_4$ composite coatings before the HT has only a single broad peak at 44.5°, which is related to a face-centered cubic (FCC) Ni (111) plane, as shown in Figure 3a. The peaks representing the C$_3$N$_4$ particles that appear in the inset of Figure 3a are not detected in the diffraction pattern of the composite coating shown in Figure 3a. This may be attributed to the low quantity of C$_3$N$_4$ and high density of Ni diffraction peaks. According to the EDX results, the NiP-C$_3$N$_4$ composite coating microstructure is a mixture of amorphous and crystalline phases. Based on the full width at half maximum (FWHM), it is found that the FWHM of the NiP and NiP-C$_3$N$_4$ composite coatings is 7.63 and 6.34, respectively. Therefore, it is concluded that the presence of C$_3$N$_4$ in the coating promotes the formation of crystalline phase; see a similar case in Ref. [26].

After HT, the as-plated NiP coating crystallized mainly as Ni$_3$P particles on the surface. As the amorphous structure is metastable, its peak decreased after HT and crystalline Ni; also, Ni$_3$P phases are formed. It is clear that the diffraction pattern of the C$_3$N$_4$ nanocomposite coating coincides with that of the NiP coating, i.e., the presence of C$_3$N$_4$ nanosheets in the coating has not affected the phase angle (peaks positions) of the coating, and new peaks have not appeared. Moreover, it can be observed that the intensity of the resulted peaks decreases dramatically for the NiP-C$_3$N$_4$ nanocomposite coating compared to C$_3$N$_4$-free one. This can explained by the decrease in the amount of deposited Ni and P as a result of presence of C$_3$N$_4$ in the coating. This is illustrated in the EDX charts shown above in Figure 1h,i, in which the Ni and P contents decreased by 3.2 wt % and 22.9 wt %, respectively.

In addition, as illustrated in Figure 1h, increasing the P(wt %) in the heat-treated NiP coating results in smaller Ni diffraction peaks.

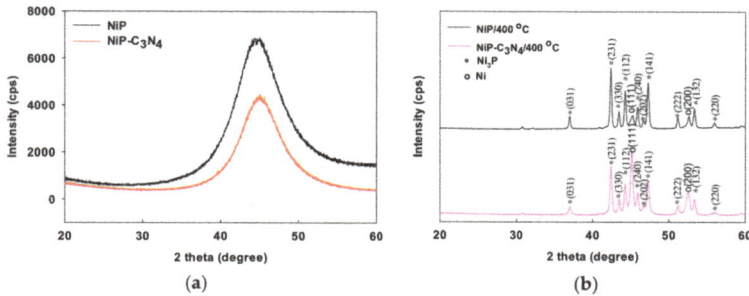

Figure 3. XRD pattern of NiP and NiP-C$_3$N$_4$ coatings (**a**) before and (**b**) after HT at 400 °C for 1 h.

3.3. Contact Angle Measurements

The surface roughness of coatings, in addition to its directly related properties, e.g., hydrophilicity and hydrophobicity, are important properties to be studied. The higher the surface roughness is, the more hydrophobic and corrosion-resistant the coating will be. Figure 4 shows the water contact angles (WCAs) of the substrate (API X120 steel), the NiP, and the NiP-C$_3$N$_4$ nanocomposite coatings before and after the HT at 400 °C for 1 h.

The WCA for steel is 86°, i.e., less than 90°, indicating its hydrophilic nature. Both as-plated NiP and NiP-C$_3$N$_4$ coatings are hydrophobic, as their contact angles are found to be 105° and 109°, respectively. After HT, a decrease in the water contact angle is observed for both NiP and the nanocomposite coatings (Figure 4d,e). Because of the recrystallization of the nickel and a phase transition that led to the change in the surface chemistry and roughness of the NiP coatings, the as-plated coatings became hydrophilic after HT [27].

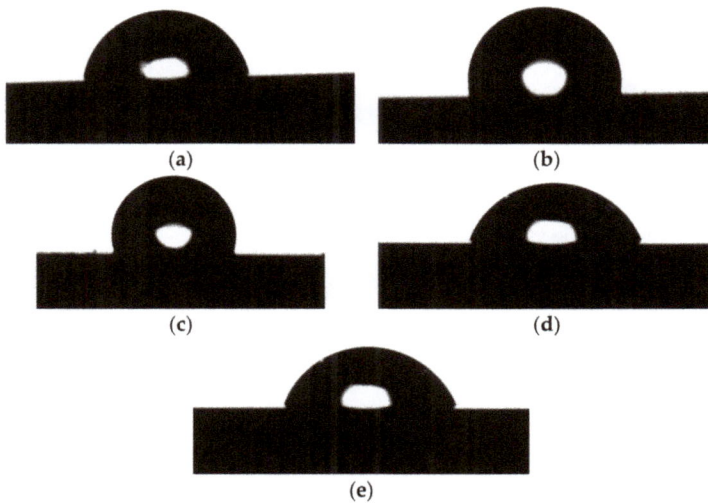

Figure 4. WCAs of coatings: (**a**) substrate: CA = 86 ± 1°; (**b**) as-plated NiP: CA = 105 ± 1°; (**c**) as-plated NiP-C$_3$N$_4$: CA = 109 ± 1°; (**d**) heat-treated NiP: CA = 70.5 ± 1°; and (**e**) heat-treated NiP-C$_3$N$_4$: CA = 72.8 ± 1°. Heat treatment was done at *T* = 400 °C for 1 h.

3.4. Surface Roughness of Coatings

The Atomic force microscopy (AFM) was used to measure the surface roughness of the NiP and NiP-C$_3$N$_4$ coatings before and after the heat treatment, as shown in Figure 5. The surface roughness of the NiP-C$_3$N$_4$ composite coating in the presence of C$_3$N$_4$ nanosheets in the NiP coating increased. The surface roughness of the as-plated NiP coating is 22 nm, whereas the NiP-C$_3$N$_4$ shows a surface roughness of approximately 43 nm. This proves the increased hydrophobicity of the NiP-C$_3$N$_4$ coating relative to that of the NiP coating. It is obvious that the surface roughness of the NiP coating and the C$_3$N$_4$ composite coating after the heat treatment decreased by about half of their initial values. This was attributed to the recrystallization of the coatings and the formation of the Ni$_3$P and Ni crystals that are more stable and show ordered structures. Therefore, the roughness of the NiP-C$_3$N$_4$ composite coating is always higher than that of the NiP coating, even after reduction of the roughness upon the heat treatment.

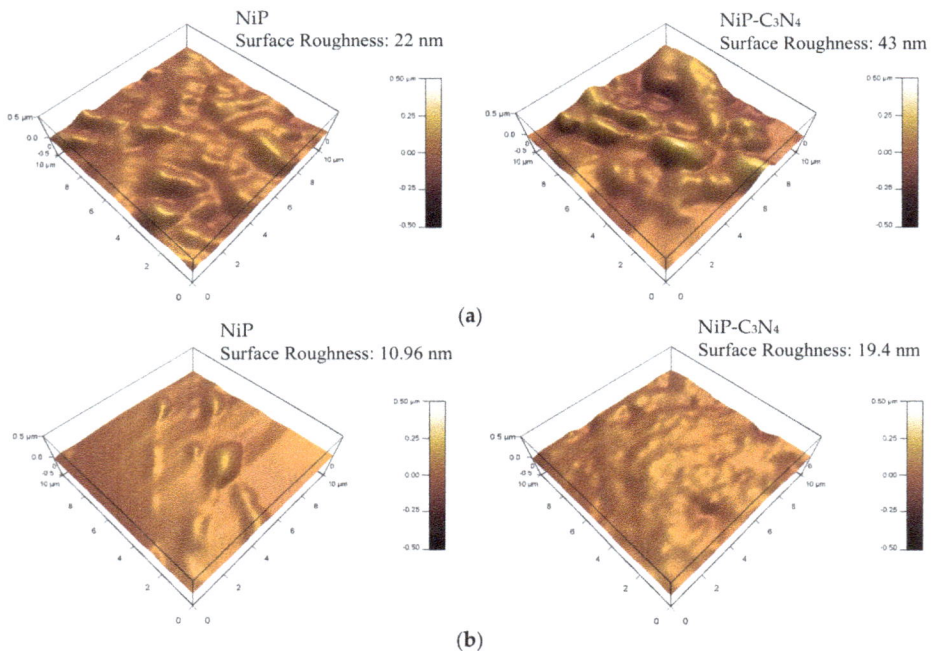

Figure 5. AFM images of electroless NiP and NiP-C$_3$N$_4$ nanocomposite coatings (**a**) before and (**b**) after heat treatment at 400 °C for 1 h.

3.5. Microhardness Measurements

The microhardness of the NiP and NiP nanocomposite coatings before and after the heat treatment were performed in addition to that of the C-steel substrate for comparison, as illustrated in Figure 6. The measured microhardness of the substrate is approximately 166 HV$_{200}$. The electroless deposition of the NiP and NiP-C$_3$N$_4$ leads to an increase in the microhardness to 406 and 645 HV$_{200}$, respectively. The increase in the microhardness upon the dispersion of the C$_3$N$_4$ nanosheets is attributed to the dispersion hardening effect caused via the incorporation of the nanoparticles into the composite coatings. As reported in [28,29], the uniform distribution of the nanoparticles in the matrix could restrain the growth of the alloy grains and the plastic deformation of the coating, leading to the stabilization of the dislocation and thus increasing the microhardness.

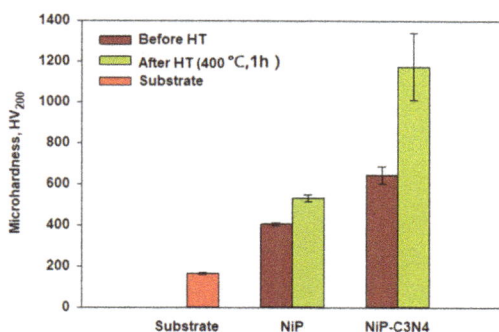

Figure 6. The microhardness of substrate, NiP coating, and NiP-C3N4 composite coating before and after heat treatment at 400 °C for 1 h.

Upon heat treatment of NiP and NiP-C$_3$N$_4$ nanocomposites, the microhardness was increased significantly to 780 and 1175 HV$_{200}$, respectively. At that temperature (400 °C), the P atoms are forced to adapt to the crystal structure of the Ni. This adaptation resulted in the formation of a coherent relationship that leads to a distortion of the local stress field. When the Ni/P ratio gathers in a sufficient quantity, the intermetallic compound Ni$_3$P precipitates and keeps a coherent relationship with the Ni. The hardness of the Ni$_3$P is gets higher when the temperature is increased, since it is formed mainly by metallic and ionic bonds, resulting in a coherent precipitation strengthening effect and an improvement of the microhardness [30]. Regarding the NiP-C$_3$N$_4$ composite, the presence of Ni$_3$P as the hard phase in addition to the C$_3$N$_4$ provides an extra factor for increasing its microhardness after the heat treatment. Although the intensity of the Ni$_3$P peaks in the NiP-C$_3$N$_4$ XRD chart is lower than the corresponding ones in the XRD chart of the C$_3$N$_4$—free coating, the hardness is higher in the case of NiP-C$_3$N$_4$ coating, indicating that the C$_3$N$_4$ compensated for the decrease in the Ni$_3$P content, which is an advantage for the new coating. In addition, it worth mentioning that the thickness of the coating in case of the NiP-C$_3$N$_4$ coating is 30% less than that in case of the NiP one.

3.6. Corrosion Measurements

3.6.1. Electrochemical Impedance Spectroscopy (EIS)

Figure 7a,b show the Bode and the phase angle plots of the EIS spectra that are measured at open circuit potential for the substrate with the as-plated and heat-treated NiP, as well as NiP-C$_3$N$_4$ nanocomposite coatings immersed in a 3.5 wt % NaCl solution at the room temperature. The larger the value of the |Z| at low frequencies is, the better the corrosion protection properties of the coating will be [31]. Inspection of Figure 7a shows that the values of |Z| at 0.01 Hz for the as-plated NiP and NiP-C$_3$N$_4$ nanocomposite coatings, as well as for the heat-treated coatings, are much higher than that of the substrate, confirming the corrosion protection properties of both coatings before and after the heat treatment. The high corrosion resistance is attributed to the presence of phosphorus [32]. Generally, for a Ni-based coating, when the nickel starts to dissolve in the corrosive media, the phosphorus starts to react with water to form a film of adsorbed hypophosphite anions, preventing further hydration of the nickel. Consequently, the corrosion resistance of the coating is increased [33]. Moreover, the as-plated NiP-C$_3$N$_4$ composite coating offers higher corrosion protection ability compared to that of the as-plated NiP coating despite the smaller phosphorous content in the former, as seen from the EDX results. This finding indicates the strong protective ability of the C$_3$N$_4$ nanosheets that enhance the polarization resistance of the NiP nanocomposite coating in 3.5 wt % NaCl solution to reach a maximum value of 9225 Ω·cm^2, as shown in Table 2. This can be attributed, as mentioned above, to the uniform distribution of C$_3$N$_4$ nanosheets throughout the coating, which support forming a more compact structure, blocking the defects in the NiP coating, inhibiting the diffusion of the chloride ions to the substrate, and enhancing the corrosion resistance as shown in a similar situation in [10].

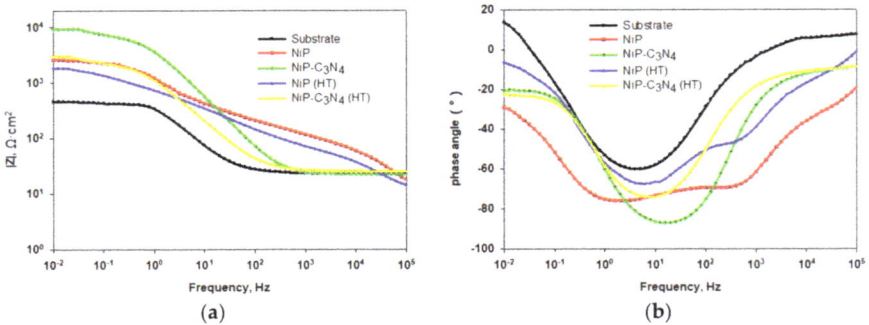

Figure 7. (a) Bode and (b) phase angle plots of the substrate, electroless as-plated NiP and NiP-C$_3$N$_4$ coatings with and without heat treatment at 400 °C for 1 h in 3.5 wt % NaCl solution at room temperature.

In addition, in Figure 7a, the Bode plots of the as-plated NiP and NiP-C$_3$N$_4$ nanocomposite coatings display different shapes in the recorded frequency regions, illustrating that different fundamental processes occur on their surfaces. This behavior is more conspicuous in the plots of the phase angle versus frequency (Figure 7b). The Bode and phase angle plots for the NiP coating show two broad peaks at the analyzed frequency range, which confirms two-time constants behavior. The first relaxation process is related to the coating layer that can be checked at the higher and intermediate frequencies, whereas the second relaxation process is observed at the lower frequencies and represents the electrochemical behavior at the interface of the substrate and the coating [34]. The Bode and phase angle plots for the NiP-C$_3$N$_4$ nanocomposite coating show one time-constant behavior, as shown in Figure 7b.

The equivalent circuits that are used in analyzing the EIS measured spectra for both the as-plated NiP and NiP-C$_3$N$_4$ coatings with and without heat treatment are depicted in Figure 8a,b, respectively. The circuit in Figure 8a includes the solution resistance (R_s), the high frequency time constant ($R_1 \cdot CPE_1$), and the low frequency time constant ($R_2 \cdot CPE_2$). The high frequency time constant ($R_1 \cdot CPE_1$) corresponds to the areas covered with the coating and can be represented by the coating admittance (CPE$_{coat}$) and the pore resistance (R_{po}). The low frequency time constant is assigned to the polarization resistance (R_p) and the admittance associated with the double layer capacitance (CPE$_{dl}$). The equivalent circuit in Figure 8b consists of the solution resistance, the double layer capacity, the polarization resistance, and the Warburg diffusion element (W). The electrochemical parameters derived from fitting the measured data using the equivalent circuits are listed in Table 2. As is clearly shown in this table, the increased polarization resistance is related to the presence of the C$_3$N$_4$ nanosheets. Moreover, the NiP-C$_3$N$_4$ composite coating has the lowest double layer capacitance (39 μF·cm^{-2}·s^{-n}) and the higher value of n (0.9) compared to those of the NiP coating. Taken together, these characteristics lead to the superior protection efficiency of the composite coating reaching as high as 95%. The protection efficiency of the NiP coating is approximately 70.9%, which is less than that of the composite due to its porosity, which allows the aggressive chloride ions to diffuse into the substrate.

After heat treatment, the polarization resistances of both heat-treated NiP and NiP-C$_3$N$_4$ are decreased compared to the corresponding polarization resistances for the as-plated coatings, but are still much greater than that of the substrate. The protective ability of the as-plated NiP and NiP-C$_3$N$_4$ composite coatings decreases after the heat treatment by approximately 11% and 9.5%, respectively, as shown in Table 2. However, the protection efficiency of the heat-treated NiP-C$_3$N$_4$ composite coating is still higher by 13% and 4% than that of the heat-treated and as-plated NiP coatings, respectively. The decrease in the corrosion protection observed for both heat-treated coatings is due to the formation of nickel phosphide (Ni_3P) that reduces the phosphorus content of the remaining material and transforms the coating from amorphous to crystalline. Previous work has shown that the amorphous alloys have better corrosion resistance than their corresponding crystalline due to the formation of glassy films that passivate their surfaces [33].

The n values for both NiP and NiP-C_3N_4 coatings before and after the heat treatment lie between 0.6 and 0.9 (Table 2). This indicates that the system is far from the ideal capacitive behavior. The deviation from the ideal capacitive behavior is related to the inhomogeneity of the coating surface attributed to the roughness and surface porosity of the coating. According to the obtained values of the CPE_{dl} presented in Table 2, the NiP-C_3N_4 composite coating before and after the heat treatment has the most homogeneous surface with a lower porosity compared to that of the NiP coating. Consequently, a dense NiP-C_3N_4 composite coating is formed on the substrate that is slightly affected by high temperatures.

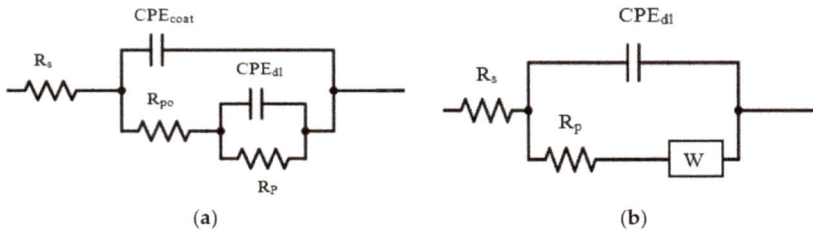

Figure 8. Equivalent electric circuits for (**a**) as-plated and heat-treated NiP coating and (**b**) as-plated and heat-treated NiP-C_3N_4 composite coating, in 3.5 wt % NaCl solution.

Table 2. Electrochemical parameters obtained by fitting the measured data shown in Figure 7 using the equivalent circuits shown in Figure 8 of the substrate, NiP, and NiP-C_3N_4 composite coatings with and without heat treatment.

Type of Coating	R_s ($\Omega \cdot cm^2$)	R_{po} ($\Omega \cdot cm^2$)	CPE_{coat} ($\mu F \cdot cm^{-2} \cdot s^{-n}$)	R_p ($\Omega \cdot cm^2$)	CPE_{dl} ($\mu F \cdot cm^{-2} \cdot s^{-n}$)	W ($S \cdot s^{1/2}$)	n	IE (%)
Substrate	22.4	–	–	445	526.5	–	–	–
NiP	17.6	167.50	28.4	2336	294	–	0.7	81
NiP-C_3N_4	21.9	–	–	9225	39	1.021×10^{-3}	0.9	95
NiP(HT)	15.6	16.9	35	1597	324.6	–	0.65	72
NiP-C_3N_4(HT)	23.4	–	–	2990	170.00	3.879×10^{-3}	0.8	85

3.6.2. Tafel Analysis

The potentiodynamic polarization curves for the substrate and electroless NiP, as well as NiP-C_3N_4 coatings before and after the heat treatment in 3.5 wt % NaCl solution at room temperature, are shown in Figure 9. The electrochemical parameters (corrosion potential (E_{corr}), corrosion current density (i_{corr}), cathodic and anodic Tafel slopes, and corrosion inhibition efficiency IE) are presented in Table 3. E_{corr} of the NiP coating (-542 mV) is shifted significantly in the anodic direction compared to the C-steel, and that of the NiP-C_3N_4 is even more anodically shifted to -309 mV. i_{corr} of the substrate, NiP, and NiP-C_3N_4 composite coatings are 21.4, 6.9, and 1.8 $\mu A \cdot cm^{-2}$, respectively. The decrease in the i_{corr} of the NiP-C_3N_4 composite coating compared to that of the NiP coating reveals a better corrosion resistance for the new composite coating. After heat treatment, E_{corr} of the NiP and the NiP-C_3N_4 composite coatings are shifted in the cathodic direction. In addition, the i_{corr} values of the NiP and NiP-C_3N_4 coatings are slightly increased compared to those before the heat treatment. This is due to the formation of crystalline Ni_3P. Consequently, an increase in i_{corr} results in a decrease in the IE to

60% and 82% for the heat-treated NiP and NiP-C$_3$N$_4$ coatings, respectively. It is worth mentioning that the *IE* of the heat-treated NiP-C$_3$N$_4$ composite coating is still higher than those of the NiP coating before and after the heat treatment. This is attributed to the existence of C$_3$N$_4$, which enhances the corrosion resistance by forming a more compact composite structure via uniform dispersion into the NiP matrix (confirmed by EDX mapping) and blocking the defects in the NiP coating [29].

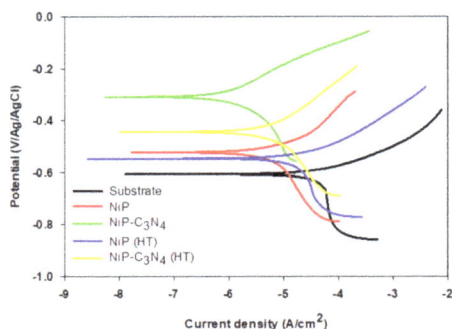

Figure 9. Polarization curves for the substrate, electroless NiP, and NiP-C$_3$N$_4$ deposits, with and without heat treatment in 3.5 wt % NaCl solution at room temperature. The scan rate is 0.167 mV·s^{-1}.

Table 3. Electrochemical parameters of different coatings before and after heat treatment derived from polarization curves shown in Figure 9.

Type of Coating	E_{corr} (mV)	i_{corr} (μA·cm^{-2})	b_a (V/decade)	b_c (V/decade)	Corrosion Rate (mpy)	IE (%)
Substrate	−607	21.4	0.1	0.17	3.55	–
NiP	−542	6.9	0.09	0.15	2.91	71.4
NiP-C$_3$N$_4$	−309	1.8	0.15	0.12	1.4	91.5
NiP(HT)	−546	8.4	0.04	0.11	6.5	60.4
NiP-C$_3$N$_4$(HT)	−444	3.8	0.11	0.12	2.26	82

4. Conclusions

The electroless deposition of the NiP-C$_3$N$_4$ nanocomposite coating is successfully achieved by sonicating the C$_3$N$_4$ nanosheets in the electroless NiP bath under the same conditions used for NiP electroless deposition. The morphology, structure, roughness, wettability, hardness, and corrosion resistance of the novel electroless NiP-C$_3$N$_4$ nanocomposite coating in comparison with the conventional NiP coating before and after heat treatment show superior properties of the new nanocomposite compared to the NiP alloy. The structure of the as-deposited NiP coating is amorphous, whereas that of the as-deposited NiP-C$_3$N$_4$ composite coating is crystalline-amorphous. The microstructure of the NiP coating is affected by the existence of the C$_3$N$_4$ nanosheets, which are uniformly spread within the NiP matrix. The existence of the C$_3$N$_4$ nanosheets and the heat treatment significantly enhances the microhardness of the NiP coating. The orderly presence of the C$_3$N$_4$ nanosheets in the coating led to the increase in the protection efficiency of the as-plated composite coating in a 3.5 wt % NaCl solution to 95% based on EIS results. After heat treatment, the formation of the crystalline Ni$_3$P phase slightly decreased the corrosion resistance of both the NiP and NiP-C$_3$N$_4$ nanocomposite coatings. However, the NiP-C$_3$N$_4$ nanocomposite still shows a corrosion protection efficiency that is higher than that of the NiP even before heat treatment.

Acknowledgments: This work was supported by the NPRP grant # NPRP8-1212-2-499 from the Qatar National Research Fund (a member of Qatar Foundation). The statements made herein are solely the responsibility of the authors.

Author Contributions: Eman M. Fayyad, Aboubakr M. Abdullah and Mohammad K. Hassan conceived and designed the experiments; Eman M. Fayyad performed the experiments; Eman M. Fayyad and Aboubakr M. Abdullah analyzed the data; Chuhong Wang, George Jarjoura and Zoheir Farhat contributed materials and revised the paper; Eman M. Fayyad wrote the first draft of the manuscript and Aboubakr M. Abdullah, Mohammad K. Hassan and Adel M. Mohamed revised the paper.

Conflicts of Interest: The authors declare no conflict of interest.

References

1. HariKrishnan, K.; John, S.; Srinivasan, K.N.; Praveen, J.; Ganesan, M.; Kavimani, P.M. An overall aspect of electroless Ni-P depositions—A review article. *Metall. Mater. Trans. A* **2006**, *37*, 1917–1926. [CrossRef]

2. Balaraju, J.N.; Narayana, T.S.N.S.; Seshadri, S.K. Electroless Ni-P composite coatings. *J. Appl. Electrochem.* **2003**, *33*, 807–816. [CrossRef]

3. Agarwala, R.C.; Agarwala, V.; Sharma, R. Electroless Ni-P based nano coating technology—A review. *Synth. React. Inorg. Met.-Org. Chem.* **2006**, *36*, 493–515. [CrossRef]

4. Makkar, P.; Agarwala, R.C.; Agarwala, V. Studies on electroless coatings at IIT Roorkee—A brief review. *Mater. Sci. Forum* **2013**, *33*, 275–288. [CrossRef]

5. Sahoo, P.; Das, S.K. Tribology of electroless nickel coatings—A review. *Mater. Des.* **2011**, *32*, 1760–1775. [CrossRef]

6. Sharma, S.B.; Agarwala, R.C.; Agarwala, V.; Ray, S. Dry sliding wear and friction behavior of Ni-P-ZrO$_2$-Al$_2$O$_3$ composite electroless coatings on aluminium. *Mater. Manuf. Processes* **2002**, *17*, 637–649. [CrossRef]

7. Srinivasan, K.N.; Thangavelu, P. Electroless deposition of Ni-P composite coatings containing kaolin nanoparticles. *Trans. Inst. Met. Finish.* **2012**, *90*, 105–112. [CrossRef]

8. Hu, J.; Fang, L.; Zhong, P. Effect of Reinforcement Particle Size on Fabrication and Properties of Composite Coatings. *Mater. Manuf. Processes* **2013**, *28*, 1294–1300. [CrossRef]

9. Sharma, A.; Singh, A.K. Electroless Ni-P and Ni-P-Al$_2$O$_3$ Nanocomposite Coatings and Their Corrosion and Wear Resistance. *J. Mater. Eng. Perform.* **2013**, *22*, 176–183. [CrossRef]

10. Zhou, H.-M.; Jia, Y.; Li, J.; Yao, S.-H. Corrosion and wear resistance behaviors of electroless Ni-Cu-P-TiN composite coating. *Rare Met.* **2016**, 1–6. [CrossRef]

11. Rezagholizadeh, M.; Ghaderi, M.; Heidary, A.; Vaghefi, S.M.M. Electroless Ni-P/Ni-B-B$_4$C duplex composite coatings for improving the corrosion and tribological behavior of Ck$_{45}$ steel. *Prot. Met. Phys. Chem. Surf.* **2015**, *51*, 234–239. [CrossRef]

12. Yang, Y.; Chen, W.; Zhou, C.; Xu, H.; Gao, W. Fabrication and characterization of electroless Ni-P-ZrO$_2$ nano-composite coatings. *Appl. Nanosci.* **2011**, *1*, 19–26. [CrossRef]

13. Soleimani, R.; Mahboubi, F.; Kazemi, M.; Arman, S.Y. Corrosion and tribological behaviour of electroless Ni-P/nano-SiC composite coating on aluminium 6061. *Surf. Eng.* **2015**, *31*, 714–721. [CrossRef]

14. Novakovic, J.; Vassiliou, P.; Samara, K.; Argyropoulos, T. Electroless Ni-P-TiO$_2$ composite coatings: Their production and properties. *Surf. Coat. Technol.* **2004**, *201*, 895–901. [CrossRef]

15. Xu, S.; Hu, X.; Yang, Y.; Chen, Z.; Chan, Y.C. Effect of carbon nanotubes and their dispersion on electroless Ni-P under bump metallization for lead-free solder interconnection. *J. Mater. Sci. Mater. Electron.* **2014**, *25*, 2682–2691. [CrossRef]

16. Liu, S.; Bian, X.; Liu, J.; Yang, C.; Zhao, X.; Fan, J.; Zhang, K.; Bai, Y.; Xu, H.; Liu, Y. Structure and properties of Ni-P-graphite (C$_g$)-TiO$_2$ composite coating. *Surf. Eng.* **2015**, *31*, 420–426. [CrossRef]

17. Ashassi-Sorkhabi, H.; Es'haghi, M. Corrosion resistance enhancement of electroless Ni-P coating by incorporation of ultrasonically dispersed diamond nanoparticles. *Corros. Sci.* **2013**, *77*, 185–193. [CrossRef]

18. Liu, A.Y.; Cohen, M.L. Prediction of new low compressibility solids. *Science* **1989**, *245*, 841. [CrossRef] [PubMed]

19. Liu, A.Y.; Cohen, M.L. Structural properties and electronic structure of low-compressibility materials: β-Si$_3$N$_4$ and hypothetical β-C$_3$N$_4$. *Phys. Rev. B* **1990**, *41*, 10727. [CrossRef]

20. Zhang, Z.; Guo, H.; Xu, Y.; Zhang, W.; Fan, X. Corrosion resistance studies on α-C$_3$N$_4$ thin films deposited on pure iron by plasma-enhanced chemical vapor deposition. *J. Mater. Sci. Lett.* **1999**, *18*, 685–687. [CrossRef]

21. Xu, H.; Yanga, Z.; Li, M.K.; Shi, Y.L.; Huang, Y.; Li, H.L. Synthesis and properties of electroless Ni-P-nanometer diamond composite coatings. *Surf. Coat. Technol.* **2005**, *191*, 161–165. [CrossRef]

22. Al-Kandari, H.; Abdullah, A.M.; Ahmad, Y.H.; Al-Kandari, S.; AlQaradawi, S.Y.; Mohamed, A.M. An efficient eco advanced oxidation process for phenol mineralization using a 2D/3D nanocomposite photocatalyst and visible light irradiations. *Sci. Rep.* **2017**, *7*, 9898. [CrossRef] [PubMed]

23. Sudagar, J.; Lian, J.; Sha, W. Electroless nickel, alloy, composite and nano coatings—A critical review. *J. Alloy. Compd.* **2013**, *571*, 183–204. [CrossRef]

24. Rajabalizadeh, Z.; Seifzadeh, D. The effect of copper ion on microstructure, plating rate and anticorrosive performance of electroless Ni-P coating on AZ61 magnesium alloy. *Prot. Met. Phys. Chem. Surf.* **2014**, *50*, 516–523. [CrossRef]

25. Afroukhteh, S.; Dehghaniann, C.; Emamy, M. Preparation of electroless Ni-P composite coatings containing nano-scattered alumina in presence of polymeric surfactant. *Prog. Nat. Sci. Mater. Int.* **2012**, *22*, 318–325. [CrossRef]

26. Zou, Y.; Cheng, Y.; Cheng, L.; Liu, W. Effect of Tin Addition on the Properties of Electroless Ni-P-Sn Ternary Deposits. *Mater. Trans.* **2010**, *51*, 277–281. [CrossRef]

27. Karthikeyan, S.; Vijayaraghavan, L. Investigation of the surface properties of heat-treated electroless Ni-P coating. *Trans. IMF* **2016**, *94*, 265–273. [CrossRef]

28. Balaraju, J.N.; Kalavati; Rajam, K.S. Influence of particle size on the microstructure, hardness and corrosion resistance of electroless Ni-P-Al$_2$O$_3$ composite coatings. *Surf. Coat. Technol.* **2006**, *200*, 3933–3941. [CrossRef]

29. Aal, A.A.; El-Sheikh, S.M.; Ahmed, Y.M.Z. Electrodeposited composite coatings of Ni-W-P with nano-sized rod and spherical-shaped SiC particles. *Mater. Res. Bull.* **2009**, *44*, 151–159. [CrossRef]

30. Liu, B.; Liu, L.; Liu, X. Effects of carbon nanotubes on hardness and internal stress in Ni-P coatings. *Surf. Eng.* **2013**, *29*, 507–510. [CrossRef]

31. Sharma, S.K. *Green Corrosion Chemistry and Engineering*; Wiley-VCH: Weinheim, Germany, 2012.

32. MafiIman, R.; Dehghanian, C. Studying the effect of the addition of TiN nanoparticles to NiP electroless coatings. *Appl. Surf. Sci.* **2011**, *258*, 1876–1880.

33. Hu, J.; Fang, L.; Liao, X.-L.; Shi, L.-T. Influences of different reinforcement particles on performances of electroless composites. *Surf. Eng.* **2017**, *33*, 362–386. [CrossRef]

34. Balaraju, J.N.; EzhilSelvi, V.; Rajam, K.S. Electrochemical behavior of low phosphorous electroless Ni-P-Si$_3$N$_4$ composite coatings. *Mater. Chem. Phys.* **2010**, *120*, 546–551. [CrossRef]

MDPI

St. Alban-Anlage 66

4052 Basel

Switzerland

Tel. +41 61 683 77 34

Fax +41 61 302 89 18

www.mdpi.com

Coatings Editorial Office

E-mail: coatings@mdpi.com

www.mdpi.com/journal/coatings

www.ingramcontent.com/pod-product-compliance
Lightning Source LLC
Chambersburg PA
CBHW051910210326
41597CB00033B/6095